Lecture Notes in Mathematics

Edited by A. Dold, Heidelberg and B. Eckmann, Zürich

353

Proceedings of the Conference on Orders, Group Rings and Related Topics

Organized by John S. Hsia, Manohar L. Madan and Thomas G. Rall
Ohio State University, Columbus, OH/USA

Springer-Verlag
Berlin · Heidelberg · New York 1973

AMS Subject Classifications (1970): 10 C 05, 12 A 90, 13 D 15, 14 G 10, 15 A 63, 16 A 18, 16 A 26, 16 A 36
16 A 46, 16 A 50, 16 A 52, 16 A 54, 16 A 64, 18 F 25, 20 C 05, 20 C 10
20 K 15, 20 K 20, 20 K 45

ISBN 3-540-06518-0 Springer-Verlag Berlin · Heidelberg · New York
ISBN 0-387-06518-0 Springer-Verlag New York · Heidelberg · Berlin

Offsetdruck: Julius Beltz, Hemsbach/Bergstr.

Dedicated to

Professor Hans J. Zassenhaus

on the occasion of his sixtieth birthday

FOREWORD

It gives me great pleasure to extend my warm good wishes and those of my colleagues to our distinguished colleague Professor Hans Zassenhaus upon the occasion of his sixtieth birthday.

Professor Zassenhaus has contributed important ideas and methods to many and diverse branches of mathematics. Among these the range of ideas concerned with Orders, Group Rings and Related Topics have held his attention for many years. His fundamental contributions to these fields have been known, appreciated and used very widely. It was thought appropriate to mark Professor Zassenhaus' sixtieth birthday by bringing together a group of fellow mathematicians who share his interest in Orders and Group Rings for a discussion of the progress to date and the prospects for the future of this difficult and important field of mathematical endeavor.

The conference which marked Professor Zassenhaus' birthday brought together a large number of able and accomplished mathematicians and resulted in much stimulating discussion. It would be difficult to describe fully the exciting and invigorating atmosphere present in a gathering of a group of creative people representing a broad spectrum of experience and of generations, who nevertheless share many interests in common. We have been able to put together a record of formal presentations which we are happy to offer in the present volume of essays.

Arnold E. Ross

PREFACE

This volume records the talks given at the Conference on Orders, Group Rings and Related Topics held at Ohio State in May, 1972, to honor our distinguished colleague Hans Zassenhaus whose contributions to mathematics in the theory of orders are widely known and appreciated. Our original intention, to convene an international gathering of mathematicians who have worked in this or some related area, had to be scaled down when it became clear that the funds available were not sufficient to support such an undertaking. None the less, as the reader examines the contents, we believe that he will share our feeling that it presents an interesting picture of areas of current interest in mathematics, a refreshing blend of expository with technical material and that the meeting is to be counted a success.

The planning and organization involved many individuals and it is appropriate that mention of their efforts be made here.

Professor Arnold E. Ross was instrumental in bringing the conference into being. From inception he encouraged its planning and when we were unable to obtain outside support, he managed to find resources for its funding. His efforts are deeply appreciated.

We thank the authors for their co-operation in the preparation of this report. Their original manuscripts were carefully done and in our hands by the agreed time. They showed great patience when various matters prevented us from holding to our original schedule.

At a meeting where current directions and new ideas for mathematical research were to be discussed, we felt it important for the forthcoming generation of mathematicians to have an opportunity to attend and participate. The Ohio State Graduate School made available money which permitted us to provide some measure of support for graduate students and we are grateful to Deans Arliss Roaden and Elmer Baumer for this.

Any conference requires attention to a large number of details which are unexciting but necessary for the smooth flow of events. Responsibility for these matters fell upon the organizing committee. As designated chairman of that committee, I want to express my gratitude to its members, John Hsia and Manohar Madan, for their service. For his major role in the preparation of this volume John deserves special commendation. The committee also wishes to express its thanks to Professor Zassenhaus for his advice and assistance in matters relating to the conference.

The final typing was done by Miss Dodie Huffman. She carried out that responsibility with skill and dedication and she has our deepest thanks for her fine work.

Finally, we are grateful to Springer Verlag, to the editors of the Lecture Note Series Professors Albrecht Dold and Beno Eckmann and to Dr. Klaus Peters and Bernd Grossmann for providing this opportunity to present the record of the conference to the general mathematical community.

T. Ralley

Table of Contents

LIST OF PARTICIPANTS

Allen, Harry
Ohio State University

Auslander, Maurice
Brandeis University

Azumaya, Goro
Indiana University

Bass, Hyman
Columbia University

Borror, Jeffry A.
Case-Western Reserve University

Brown, Robert
Ohio State University

Carlson, Jon F.
University of Georgia

Chang, Kuang-I
Ohio State University

Chang, Morgan F. H.
Columbia University

Chin, W.
Ohio State University

Chittenden, Charles
Ohio State University

Cliff, Gerald
University of Illinois

Coleman, Donald
University of Kentucky

Cunningham, Joel
University of Kentucky

Dennis, Keith
Cornell University

Divis, Bohuslav
Ohio State University

Eldridge, Klaus E.
Ohio University

Elkins, Bryce
Ohio State University

Endo, Lawrence
University of Illinois

Ferrar, Joe
Ohio State University

Fossum, T. V.
University of Illinois

Galovich, Steve
Carleton College

Cillam, John D.
Ohio University

Glover, Henry
Ohio State University

Green, Edward L.
Brandeis University

Gustafson, William H.
Indiana University

Hannula, Thomas
University of Maine

Hill, Walter
Princeton University

Hsia, John
Ohio State University

Hughes, Ian
Queen's University

Jain, S. K.
Ohio University

James, Donald G.
Penn State University

Johnson, Robert P.
Ohio State University

Kimmel, David
Ohio State University

Klasa, J.
McGill University

Klasa, S.
McGill University

Liang, Joseph
University of South Florida

Madan, M.
Ohio State University

Martin, Robert
Columbia University

McCulloh, Leon
University of Illinois

Merklen, Hector A.
Ohio State University

Miller, Len
Ohio State University

Mislin, Guido
Ohio State University

Moreno, Carlos S.
University of Illinois

Osterburg, James
University of Cincinnati

Peterson, Roger
Ohio State University

Pollack, David
Queen's University

Queen, Clifford S.
Ohio State University

Ralley, T.
Ohio State University

Reiner, Irving
University of Illinois

Reiten, Idun
Brandeis University

Ross, A. E.
Ohio State University

Sallee, William
University of Kentucky

Santa Pietro, John
Rutgers University

Sehgal, S. K.
University of Alberta

Siu, Man Keung
Columbia University

Swan, Richard G.
University of Chicago

Symonds, Robin G.
Ohio University

Ullom, Stephen
University of Illinois

Upton, John
University of Maryland

Wang, Chang-Yean
University of Illinois

Ware, Roger
Northwestern University

Wei, Chou-Hsiang
Queen's University

Yamada, Toshihiko
Queen's University

Yang, John
Ohio State University

Zassenhaus, Hans
Ohio State University

Zilber, Joseph
Ohio State University

THE NON-TRIVIALITY OF $SK_1(\mathbb{Z}\pi)$

Roger C. Alperin[1], R. Keith Dennis[2] and Michael R. Stein[3]

Rice University, Cornell University and Northwestern University

In this note we give examples of finite abelian groups π for which $SK_1(\mathbb{Z}\pi)$ is non-trivial. Although this note does not correspond to a talk given at this conference, it reports on work which was inspired by Bass' talk at the conference [6] in which he raised this question. Although the problem of computing $SK_1(\mathbb{Z}\pi)$ is of interest to topologists, these groups have been determined in relatively few cases (cf. [4], [5], [11], [12]). In the following discussion we outline a method which, in principle, allows one to compute $SK_1(\mathbb{Z}\pi)$ for any finite abelian group π. In the case where π is an elementary abelian p-group, a precise answer is given. Details of the proofs will appear elsewhere.

For any associative ring with unit, $GL(R)$ denotes the direct limit of the general linear groups $GL(n,R)$. $K_1(R)$ is $GL(R)$ abelianized. There is always a homomorphism $U(R) \to K_1(R)$ and in the case that R is commutative this homomorphism is split by the determinant homomorphism $K_1(R) \to U(R)$ [4, p. 229]. This yields the decomposition $K_1(R) \approx U(R) \oplus SK_1(R)$ where $SK_1(R)$ denotes the kernel of the determinant homomorphism.

If R is the group ring $\mathbb{Z}\pi$, the Whitehead group of π, $Wh(\pi)$, is defined by $K_1(\mathbb{Z}\pi)/\{\pm\pi\}$ where $\{\pm\pi\}$ denotes the image of $\pm\pi$ under the map $U(\mathbb{Z}\pi) \to K_1(\mathbb{Z}\pi)$. When π is a finite group, $Wh(\pi)$ is a finitely generated abelian group whose rank was computed by Bass [4, Theorem 7.5, p. 625]. For a

---

[1] Supported by an NDEA fellowship. Part of this research constitutes the first author's doctoral dissertation at Rice University under the direction of S. M. Gersten.

[2] Supported by NSF-GP-25600.

[3] Supported by NSF-GP-28915.

finite group π the only torsion elements in $U(\mathbb{Z}\pi)$ are of the form $\pm g$, $g \in \pi$, [10] and hence in the case of a finite abelian group, $Wh(\pi) \approx SK_1(\mathbb{Z}\pi) \oplus F$ where F is a finitely generated free abelian group. It is known that $SK_1(\mathbb{Z}\pi)$ is a finite group [4, Chapter XI, Theorem 7.5a] and in all previous cases where $SK_1(\mathbb{Z}\pi)$ has been computed, it is trivial [4, Chapter XI, Proposition 7.3 and Remark on p. 624].

Let $A \to B = \Pi B_i$ be an inclusion of rings with the maps $A \to B_i$ surjective. If $\underline{c}$ is a 2-sided ideal of B contained in A, then there is a Cartesian square

$$\begin{array}{ccc} A & \to & B \\ \downarrow & & \downarrow \\ A/\underline{c} & \to & B/\underline{c} \end{array}$$

and a theorem of Bass [4, Theorem 5.8, p. 484] allows us to extend the Mayer-Vietoris sequence of K-theory to the left:

$$K_2(B) \oplus K_2(A/\underline{c}) \to K_2(B/\underline{c}) \to K_1(A) \to K_1(B) \oplus K_1(A/\underline{c}) \to \ldots$$

If π is a finite abelian group, we apply this sequence with $A = \mathbb{Z}\pi$, B the integral closure of $\mathbb{Z}\pi$ in $\mathbb{Q}\pi$, and $\underline{c}$ the conductor of B over A. Now $K_1(B) = U(B)$ [7, Corollary 4.3] and $K_1(A/\underline{c}) = U(A/\underline{c})$ since $A/\underline{c}$ is a finite commutative ring. Thus $Ker(K_1(A) \to K_1(B) \oplus K_1(A/\underline{c})) = SK_1(A)$ and we obtain the exact sequence

$$(1) \qquad K_2(B) \oplus K_2(A/\underline{c}) \to K_2(B/\underline{c}) \to SK_1(A) \to 1.$$

If $\pi \to \pi'$ is a surjection of finite abelian groups, we obtain a commutative diagram

$$\begin{array}{ccccccc} K_2(B) \oplus K_2(A/\underline{c}) & \to & K_2(B/\underline{c}) & \to & SK_1(A) & \to & 1 \\ \downarrow & & \downarrow & & \downarrow & & \\ K_2(B') \oplus K_2(A'/\underline{c}') & \to & K_2(B'/\underline{c}') & \to & SK_1(A') & \to & 1. \end{array}$$

Now $B/\underline{c} \to B'/\underline{c}'$ is surjective and it follows that $K_2(B/\underline{c}) \to K_2(B'/\underline{c}')$ is surjective since each is generated by the symbols $\{u,v\}$ [14, Theorem 2.13] and units in $B'/\underline{c}'$ can be lifted to units in $B/\underline{c}$. This yields

THEOREM 1. <u>A surjection</u> $\pi \to \pi'$ <u>of finite abelian groups induces a surjection</u> $SK_1(\mathbb{Z}\pi) \to SK_1(\mathbb{Z}\pi')$.

Now B is the direct product of rings of cyclotomic integers B_i and $\underline{c}$ decomposes into a product of ideals $\underline{c}_i \subset B_i$ [8]. As K_2 preserves finite products, the computation of [9, Theorem 5.1] completely determines the finite group $K_2(B/\underline{c})$. In particular, an exponent for π is an exponent for $K_2(B/\underline{c})$ and hence also for $SK_1(A)$. We first determine the effect of the homomorphism $K_2(B) \to K_2(B/\underline{c})$ on exact sequence (1). This is nothing other than the homomorphism derived from the algebraic K-theory exact sequence [13, Theorem 6.2]

$$K_2(B) \to K_2(B/\underline{c}) \to SK_1(B,\underline{c}) \to 1$$

which itself is the direct product of the exact sequences

(2) $\quad K_2(B_i) \to K_2(B_i/\underline{c}_i) \to SK_1(B_i,\underline{c}_i) \to 1.$

To simplify the computation, we henceforth will assume that π is an abelian p-group. The computation of $SK_1(B_i,\underline{c}_i)$ [7, Corollary 4.3] and of $K_2(B_i/\underline{c}_i)$ [9, Theorem 5.1] together with (2) yield

(i) The map $K_2(B_i) \to K_2(B_i/\underline{c}_i)$ is zero if B_i is totally imaginary, and

(ii) the map $K_2(B_i) \to K_2(B_i/\underline{c}_i)$ is surjective if B_i is real.

The well-known result that $SK_1(\mathbb{Z}\pi)$ is trivial when π is an elementary abelian 2-group follows immediately from (ii) and exact sequence (1).

In view of (ii) above, we let B_o be the direct product of those $B_i/\underline{c}_i$ which are totally imaginary and obtain the exact sequence

$$(3) \qquad K_2(A/\underline{c}) \to K_2(B_o) \to SK_1(A) \to 1 .$$

Since π is an abelian p-group, $A/\underline{c}$ is a finite local ring and hence $K_2(A/\underline{c})$ is generated by symbols [14, Theorem 2.13]. In view of exact sequence (3), the computation of $SK_1(\mathbb{Z}\pi)$ for a finite abelian p-group is reduced to a finite computation.

At this point, as a technical convenience, the ring B_o is replaced by a homomorphic image $\overline{B}$ which is the smallest ring with the property that $K_2(B_o) \to K_2(\overline{B})$ is an isomorphism. $A/\underline{c}$ is also replaced by another local ring $\overline{A}$ whose K_2 is more amenable to computation and which has the property that

$$K_2(\overline{A}) \to K_2(\overline{B}) \to SK_1(A) \to 1$$

ramains exact.

A dévissage argument using a filtration of ideals in $\overline{A}$ together with a theorem of Stein [14, Theorem 2.5] allow us to reduce the collection of symbols generating $K_2(\overline{A})$ to a manageable number. We next apply [9, §4, Remark 2] to the computation of $K_2(\overline{B})$. This computation uses the norm residue symbol. In the case of cyclotomic fields of p^n-th roots of unity, explicit formulas for the norm residue symbol are known ([2], [3, Theorem 10, p. 164]). For an elementary abelian p-group these formulas allow us to extract from our collection of generators of $K_2(\overline{A})$ a set of symbols whose images in $K_2(\overline{B})$ independently generate the image of $K_2(\overline{A}) \to K_2(\overline{B})$. This yields the following result.

THEOREM 2. *If* p *is an odd prime and* π *is an elementary abelian* p-*group of rank* k, *then* $SK_1(\mathbb{Z}\pi)$ *is an elementary abelian* p-*group of rank*

$$\frac{p^k - 1}{p - 1} - \binom{p + k - 1}{p}.$$

In particular, $SK_1(\mathbb{Z}\pi)$ is never trivial if $k \geq 3$. In view of Theorem 1, this shows that $SK_1(\mathbb{Z}\pi)$ is non-trivial for "most" finite abelian groups π.

REFERENCES

[1] R. Alperin, Thesis, Rice University, 1973.

[2] E. Artin and H. Hasse, Die beiden Ergänzungssätze zum Reziprozitätsgesetz der ℓ^n-ten Potenzreste im Körper der ℓ^n-ten Einheitswurzeln, Abh. Math. Sem. Univ. Hamburg 6 (1928), 146-162.

[3] E. Artin and J. Tate, Class Field Theory, Benjamin, New York, 1967.

[4] H. Bass, Algebraic K-theory, Benjamin, New York, 1968.

[5] ________, The Dirichlet unit theorem, induced characters, and Whitehead groups of finite groups, Topology 4 (1966), 391-410.

[6] ________, these Proceedings.

[7] H. Bass, J. Milnor and J.-P. Serre, Solution of the congruence subgroup problem for $SL_n (n \geq 3)$ and $Sp_{2n} (n \geq 2)$, Publ. Math. IHES No. 33 (1967), 59-137.

[8] H. Bass and M. P. Murthy, Grothendieck groups and Picard groups of abelian group rings, Ann. of Math. 86 (1967), 16-73.

[9] R. K. Dennis and M. R. Stein, K_2 of discrete valuation rings, (to appear).

[10] G. Higman, The units of group rings, Proc. London Math. Soc. 46 (1940), 231-248.

[11] T.-Y. Lam, Induction theorems for Grothendieck groups and Whitehead groups of finite groups, Ann. Sci. Ec. Norm. Sup. (4) 1 (1968), 91-148.

[12] J. Milnor, Whitehead Torsion, Bull. Amer. Math. Soc. 72 (1966), 358-426.

[13] ________, Introduction to algebraic K-theory, Annals of Math. Studies No. 72, Princeton University Press, Princeton, 1971.

[14] M. R. Stein, Surjective stability in dimension 0 for K_2 and related functors, Trans. Amer. Math. Soc., (to appear).

Rice University, Houston, Texas 77001.

Cornell University, Ithaca, New York 14850.

Northwestern University, Evanston, Illinois 60201. (Mailing address for 1972-73: Institute of Mathematics, Hebrew University, Jerusalem, Israel.)

STABLE EQUIVALENCE OF ARTIN ALGEBRAS

Maurice Auslander and Idun Reiten*

Brandeis University

INTRODUCTION. An artin algebra is an Artin ring which is a finitely generated module over its center R, which is also an Artin ring. For an Artin algebra Λ we denote by $\mathrm{mod}(\Lambda)$ the category of finitely generated (left) Λ-modules. Two algebras Λ and Λ' are said to be Morita equivalent if $\mathrm{mod}(\Lambda)$ and $\mathrm{mod}(\Lambda')$ are equivalent categories.

We will consider another category associated with Λ, the projectively stable category of modules, denoted by $\underline{\mathrm{mod}(\Lambda)}$. The objects are the finitely generated Λ-modules, which we will denote by $\underline{M}$, and the morphisms are given by $\mathrm{Hom}(\underline{M},\underline{N}) = \mathrm{Hom}_\Lambda(M,N)/R(M,N)$ where $R(M,N)$ is the subgroup of $\mathrm{Hom}_\Lambda(M,N)$ consisting of the Λ-homomorphisms from M to N which factor through a projective Λ-module. Let $\mathrm{mod}_P(\Lambda)$ denote the full subcategory of $\mathrm{mod}(\Lambda)$ whose objects are the Λ-modules with no projective direct summands. Then the corresponding projectively stable category $\underline{\mathrm{mod}_P(\Lambda)}$ with morphisms as described above, is equivalent to $\underline{\mathrm{mod}(\Lambda)}$.

Analogously we define the injectively stable category $\overline{\mathrm{mod}(\Lambda)}$ associated to $\mathrm{mod}(\Lambda)$. The objects of $\overline{\mathrm{mod}(\Lambda)}$ are the same as the objects of $\mathrm{mod}(\Lambda)$, denoted by $\overline{M}$. And the morphisms are given by $\mathrm{Hom}(\overline{M},\overline{N}) = \mathrm{Hom}_\Lambda(M,N)/U(M,N)$, where $U(M,N)$ is the subgroup of $\mathrm{Hom}_\Lambda(M,N)$ consisting of the Λ-homomorphisms from M to N which factor through an injective Λ-module. And if $\mathrm{mod}_I(\Lambda)$ denotes the full subcategory of $\mathrm{mod}(\Lambda)$ whose objects are the Λ-modules with no injective direct summands, then $\overline{\mathrm{mod}_I(\Lambda)}$ and $\overline{\mathrm{mod}(\Lambda)}$ are equivalent. For

*This paper was written while the authors were partially supported by NSF GP 28486, and the second author by NAVF (Norwegian Research Council).

an Artin algebra Λ, $\underline{\mathrm{mod}}(\Lambda)$ and $\overline{\mathrm{mod}}(\Lambda)$ turn out to be equivalent categories. We will say that Λ and Λ' are <u>stably equivalent</u> if $\underline{\mathrm{mod}}(\Lambda)$ and $\underline{\mathrm{mod}}(\Lambda')$ are equivalent.

There are two easy cases where non-isomorphic algebras are stably equivalent: If Λ and Λ' are Morita equivalent, they are clearly stably equivalent. And it is also easy to see that if S is a semi-simple algebra, then Λ and $\Lambda \times S$ are stably equivalent. We ask what connection there has to be between Λ and Λ' for the algebras to be stably equivalent.

There is a natural functor $F : \mathrm{mod}_p(\Lambda) \to \underline{\mathrm{mod}}(\Lambda)$, where $F(M) = \underline{M}$, which is a representation equivalence [3] [4], i.e. F induces a one-one correspondence between the indecomposable objects of $\mathrm{mod}_p(\Lambda)$ and $\underline{\mathrm{mod}}(\Lambda)$. Hence there is a close connection between the representation theory for Λ and Λ' if they are stably equivalent. In particular, Λ is of finite representation type if and only if Λ' is.

To study the categories $\underline{\mathrm{mod}}(\Lambda)$ and $\overline{\mathrm{mod}}(\Lambda)$, we define a suitable abelian category $\hat{\mathcal{C}}_0(\Lambda)$, with enough projectives and injectives. This category will have the property that if $\mathcal{P}$ denotes the full subcategory of projective objects, $\mathcal{J}$ the full subcategory of injective objects of $\hat{\mathcal{C}}_0(\Lambda)$, there are natural equivalences of categories $\alpha : \underline{\mathrm{mod}}(\Lambda) \to \mathcal{P}$ and $\beta : \overline{\mathrm{mod}}(\Lambda) \to \mathcal{J}$. So $\hat{\mathcal{C}}_0(\Lambda)$ determines the stable category $\underline{\mathrm{mod}}(\Lambda)$, and conversely $\hat{\mathcal{C}}_0(\Lambda)$ is determined by $\underline{\mathrm{mod}}(\Lambda)$. Instead of asking directly when $\underline{\mathrm{mod}}(\Lambda)$ and $\underline{\mathrm{mod}}(\Lambda')$ are equivalent, we can ask when $\hat{\mathcal{C}}_0(\Lambda)$ and $\hat{\mathcal{C}}_0(\Lambda')$ are equivalent. To be able to use this we have to get some results on $\hat{\mathcal{C}}_0(\Lambda)$.

In Chapter I we set up our machinery, leading to the definition of $\hat{\mathcal{C}}_0(\Lambda)$. It is defined to be the full subcategory of the coherent functors $\hat{\mathcal{C}} \subset (\mathcal{C}^{op}, \mathrm{Ab})$, where $\mathcal{C} = \mathrm{mod}(\Lambda)$, which vanish on projective objects. We give a description of the projective and injective objects of $\hat{\mathcal{C}}_0(\Lambda)$, and show that $\hat{\mathcal{C}}_0(\Lambda)$ has minimal projective covers by first establishing that $\hat{\mathcal{C}}(\Lambda)$ does. We describe injective and projective resolutions for objects of $\hat{\mathcal{C}}_0(\Lambda)$, which turn out to be minimal in the first two steps, a fact which turns out to be useful in what follows.

In Chapter II we apply these results to compare properties of a Λ-module M, and $\alpha(\underline{M})$ or $\beta(\overline{M})$ in $\hat{\mathcal{C}}_0(\Lambda)$. And we use this to deduce necessary connections between two algebras which are stably equivalent.

Our main aim is to give a characterization of the algebras which are stably equivalent to hereditary algebras, and furthermore to show that each stable equivalence class has essentially one hereditary algebra (i.e. up to Morita equivalence and a semi-simple ring summand).

To handle this, we prove a structure theorem for hereditary algebras, which we apply along with some results of Chapter II to show that each stable equivalence class has essentially one hereditary algebra.

Combining results of Gabriel [7] and Yoshii [11] with results of Mitchell [10], one can classify a big class of hereditary algebras, as to whether they are of finite type or not, namely hereditary algebras which are finite products of subrings of full lower triangular matrix rings with entries in a field (division ring) k, $Tr_n(k)$. This will be discussed elsewhere. Hence it is of some interest to investigate which algebras are stably equivalent to hereditary algebras.

In Chapter IV we give the following necessary and sufficient conditions for an Artin algebra Λ with radical $\mathfrak{r}$ to be stably equivalent to an hereditary algebra:

(1) Each indecomposable submodule of an indecomposable projective module is projective or simple.

(2) For each torsionless non-projective simple module S there is an injective module E with $S \subset E/\mathfrak{r}E$.

We give examples to show that neither (1) nor (2) can be left out, and examples of rings with different global dimensions satisfying properties (1) and (2). In fact, we show that each Artin algebra with $\mathfrak{r}^2 = 0$ belongs to this class, as it is easily seen that (1) and (2) are satisfied in this case.

In the last chapter, we give a different approach to the case $\mathfrak{r}^2 = 0$. Our method here works for arbitrary Artin rings, rather than Artin algebras, so we show that each Artin ring with $\mathfrak{r}^2 = 0$ is (projectively) stably equivalent to an hereditary ring. And since stably equivalent rings have closely related representation theories, a consequence of this is that to classify Artin rings with $\mathfrak{r}^2 = 0$ of finite representation type, it is sufficient to classify the hereditary ones.

This approach also gives an easy way of constructing an hereditary ring in the same stable equivalence class as Λ, namely the triangular matrix ring $\Gamma = \begin{pmatrix} \Lambda/\mathfrak{r} & 0 \\ \mathfrak{r} & \Lambda/\mathfrak{r} \end{pmatrix}$. In fact, we show this by directly constructing an equivalence $F : \underline{\text{mod}(\Lambda)} \to \underline{\text{mod}(\Gamma)}$. This will enable us to formulate necessary and sufficient conditions for two Artin algebras Λ and Λ' with $\mathfrak{r}^2 = \mathfrak{r}'^2 = 0$, to be stably equivalent, in terms of $\mathfrak{r}$ and $\mathfrak{r}'$.

We will assume that all our Artin rings Λ have the property that $\Lambda/\mathfrak{r}$ is a product of division rings. There is no loss of generality, since any Artin ring is Morita equivalent to an Artin ring with this property. We also assume that all Λ-modules are finitely generated left Λ-modules.

These notes are a preliminary, informal report on some of these results. A more complete formal presentation will be made elsewhere.

CHAPTER I

§1. THE STABLE CATEGORIES. Let Λ be an Artin algebra, i.e. an Artin ring which is a finitely generated module over its center R. We let mod(Λ) denote the category of finitely presented Λ-modules, which here coincides with the category of finitely generated (left) Λ-modules. We want to study the category

$\underline{\mathrm{mod}(\Lambda)}$, the projectively stable category associated to Λ. The objects of $\mathrm{mod}(\Lambda)$ are the same as those of $\mathrm{mod}(\Lambda)$, i.e. the finitely generated (left) Λ-modules, denoted by $\underline{M}$. Let $R(M,N) = \{f : M \to N$, such that f factors through a projective Λ-module$\}$. $R(M,N)$ is then a subgroup of $\mathrm{Hom}_\Lambda(M,N)$, and $\mathrm{Hom}(\underline{M},\underline{N})$ is defined to be the quotient group $\mathrm{Hom}_\Lambda(M,N)/R(M,N)$. If $f \in \mathrm{Hom}(M,N)$, we denote by $\underline{f}$ its image in $\mathrm{Hom}(\underline{M},\underline{N})$. If $\underline{f} \in \mathrm{Hom}(\underline{M},\underline{N})$, $\underline{g} \in \mathrm{Hom}(\underline{N},\underline{L})$, the composition $\underline{g} \circ \underline{f} \in \mathrm{Hom}(\underline{M},\underline{L})$ is defined to be $\underline{g \circ f}$. It is easy to see that $\underline{g \circ f}$ is independent of the choice of representatives for $\underline{f}$ and $\underline{g}$. $\underline{\mathrm{mod}(\Lambda)}$ is an additive, but as we shall see, not necessarily an abelian category. If we consider the full subcategory $\mathrm{mod}_P(\Lambda)$, $\mathrm{mod}(\Lambda)$, whose objects are the finitely generated Λ-modules with no projective summands, and the corresponding projectively stable category with morphisms as described above, we get a category $\underline{\mathrm{mod}_P(\Lambda)}$ equivalent to $\underline{\mathrm{mod}(\Lambda)}$.

We will also consider associated to $\mathrm{mod}(\Lambda)$ the analogously defined injectively stable category $\overline{\mathrm{mod}(\Lambda)}$. Again the objects are the objects of $\mathrm{mod}(\Lambda)$, denoted by $\overline{M}$. Let $U(M,N) = \{f : M \to N$, such that f factors through an injective Λ-module$\}$. Then $\mathrm{Hom}(\overline{M},\overline{N}) = \mathrm{Hom}(M,N)/U(M,N)$. Similarly, if $\mathrm{mod}_I(\Lambda)$ denotes the full subcategory of $\mathrm{mod}(\Lambda)$ whose objects are the finitely generated Λ-modules with no injective summands, then $\overline{\mathrm{mod}_I(\Lambda)}$ and $\overline{\mathrm{mod}(\Lambda)}$ are equivalent.

Furthermore, the categories $\underline{\mathrm{mod}_P(\Lambda)}$ and $\overline{\mathrm{mod}_I(\Lambda)}$ are actually equivalent. For let M be an object of $\mathrm{mod}_P(\Lambda)$. Then $T(M) \in \mathrm{mod}_P(\Lambda^{op})$ (where Λ^{op} denotes the opposite ring) is defined as follows [5] : Let $P_1 \to P_0 \to M \to 0$ be a minimal projective presentation of M, and let $T(M)$ be such that the following sequence is exact: $P_0^* \to P_1^* \to T(M) \to 0$, where $P_i^* = \mathrm{Hom}_\Lambda(P_i, \Lambda)$. T induces a duality $T : \underline{\mathrm{mod}(\Lambda)} \to \underline{\mathrm{mod}(\Lambda^{op})}$ [4]. Let M be a Λ^{op}-module, and $D(M) = \mathrm{Hom}_R(M, E_R)R/\mathfrak{m}))$, where $E_R(R/\mathfrak{m})$ denotes the injective envelope of $R/\mathfrak{m}$. D gives a duality $D : \mathrm{mod}(\Lambda^{op}) \to \mathrm{mod}(\Lambda)$, which induces a duality

$D : \text{mod}_P(\Lambda^{op}) \to \text{mod}_I(\Lambda)$. It is also easy to see that we get induced a duality, also denoted by D, from $\underline{\text{mod}}(\Lambda^{op})$ to $\overline{\text{mod}}(\Lambda)$. Hence the composition $D \circ T : \underline{\text{mod}}(\Lambda) \to \underline{\text{mod}}(\Lambda^{op}) \to \overline{\text{mod}}(\Lambda)$ gives our desired equivalence. This justifies the shorter term stable equivalence which we will adopt, rather than injectively or projectively stable, when we study Artin algebras.

We want to set up machinery for studying the relationships between two algebras Λ and Λ' which are stably equivalent. We are going to study an abelian category $\hat{\mathcal{C}}_0(\Lambda)$ with enough projectives and injectives assoicated with Λ, and which contains all information about $\underline{\text{mod}}(\Lambda)$ (and $\overline{\text{mod}}(\Lambda)$). $\underline{\text{mod}}(\Lambda)$ will namely be equivalent to the full subcategory of projective objects of $\hat{\mathcal{C}}_0(\Lambda)$ and $\overline{\text{mod}}(\Lambda)$ equivalent to the full subcategory of injective objects. Before we define this category, we shall need to introduce some other concepts.

§2. VARIOUS CATEGORIES. To an additive category $\mathcal{A}$, we will associate some other categories, defined in a natural way (see [4] for more details).

Let Morph $\mathcal{A}$ denote the category whose objects are triples (A,B,f), where $f : A \to B$ is a morphism in $\mathcal{A}$, and a morphism (g,h) between two objects (A,B,f) and (A',B',f') is a pair of maps $g : A \to A'$, $h : B \to B'$ in $\mathcal{A}$, such that

$$\begin{array}{ccc} A & \xrightarrow{f} & B \\ g\downarrow & & \downarrow h \\ A' & \xrightarrow{f'} & B' \end{array}$$

commutes. We say that (g,h) is projectively trivial if there is a map $i : B \to A'$ such that $f'i = h$. And (g,h) is said to be injectively trivial if there is a map $i : B \to A'$ such that $if = g$.

Let X and Y be objects in Morph $\mathcal{A}$. Then the projectively trivial maps from X to Y form a subgroup $P(X,Y)$ of $\text{Hom}(X,Y)$ and the injectively trivial maps a subgroup $I(X,Y)$.

We can now define two new categories Mod $\mathcal{A}$ and Comod $\mathcal{A}$. The objects of Mod $\mathcal{A}$ are the same as those of Morph $\mathcal{A}$, and the morphisms are given by $\text{Hom}_{\text{Mod}\,\mathcal{A}}(X,Y) = \text{Hom}(X,Y)/P(X,Y)$. The objects of Comod $\mathcal{A}$ are also the objects of Morph $\mathcal{A}$, and the morphisms are $\mathrm{Hom}_{\mathrm{Comod}\,\mathcal{A}}(X,Y) = \mathrm{Hom}(X,Y)/I(X,Y)$.

We point out that if $\mathcal{A}$ is an additive subcategory of some abelian category $\mathcal{B}$, and if all objects of $\mathcal{A}$ are projective in $\mathcal{B}$, then there is a natural functor $F : \mathrm{Mod}\,\mathcal{A} \to \mathcal{B}$, induced by the natural full functor $F' : \mathrm{Morph}\,\mathcal{A} \to \mathcal{B}$, $F'(A,B,f) = \mathrm{Coker}\, f$, which is fully faithful. In particular, if $\mathcal{B}$ has enough projectives, and $\mathcal{A}$ consists of all projective objects, then $F : \mathrm{Mod}\,\mathcal{A} \to \mathcal{B}$ is an equivalence of categories. For example, if Λ is an Artin algebra, and $\mathcal{P}$ denotes the full subcategory of finitely generated Λ-modules, then $\mathrm{Mod}(\mathcal{P})$ and $\mathrm{mod}(\Lambda)$ are equivalent.

We shall also need the analogous observation that if $\mathcal{A}$ is an additive subcategory of an abelian category $\mathcal{B}$ consisting of injective objects, there is a natural fully faithful functor $G : \mathrm{Comod}\,\mathcal{A} \to \mathcal{B}$. Here $G(A,B,f) = \mathrm{Ker}\, f$.

If $F : \mathcal{A} \to \mathcal{B}$ is a duality, there is induced, in a natural way, a duality $F' : \mathrm{Mod}\,\mathcal{A} \to \mathrm{Comod}\,\mathcal{B}$, given by $F'(A,B,f) = (F(B),F(A),F(f))$.

To get another description of $\mathrm{Mod}\,\mathcal{A}$, we assume now that $\mathcal{A}$ is skeletally small (i.e. that the isomorphism classes of objects form a set). Consider the abelian category $(\mathcal{A}^{op},\mathrm{Ab})$ of functors from $\mathcal{A}^{op}$ to Ab. The natural functor $G' : \mathrm{Morph}\,\mathcal{A} \to (\mathcal{A}^{op},\mathrm{Ab})$, given by $G'(A,B,f) = \mathrm{Coker}((\ ,A) \xrightarrow{(\ ,f)} (\ ,B))$ induces a fully faithful functor $G : \mathrm{Mod}\,\mathcal{A} \to (\mathcal{A}^{op},\mathrm{Ab})$. We say that a functor $F : \mathcal{A}^{op} \to \mathrm{Ab}$ is coherent (see [2] and [4]) if there is an exact sequence $(\ ,A) \to (\ ,B) \to F \to 0$. We denote the full subcategory of $(\mathcal{A}^{op},\mathrm{Ab})$ of coherent functors by $\hat{\mathcal{A}}$. Then $G : \mathrm{Mod}\,\mathcal{A} \to (\mathcal{A}^{op},\mathrm{Ab})$ induces an equivalence of categories between $\mathrm{Mod}\,\mathcal{A}$ and $\hat{\mathcal{A}}$. We remark that $\mathrm{gl.\,dim}\,\hat{\mathcal{A}} \leq 2$ if $\mathcal{A}$ has kernels [4].

If A is a representation generator for the additive category $\mathcal{A}$, i.e. each object B in $\mathcal{A}$ is a summand of a finite direct sum of copies of A, there is yet another way of looking at $\mathrm{Mod}\,\mathcal{A}$ and $\hat{\mathcal{A}}$, which will be useful in what follows. There is a natural functor $H : \hat{\mathcal{A}} \to \mathrm{mod}(\mathrm{End}(A)^{op})$ (= finitely presented $\mathrm{End}(A)^{op}$-modules) given by $F \to F(A)$. $F(A)$ is an abelian group, which has a natural structure as an End(A)^{op}-module. H is an equivalence of categories. In the cases we will need End(A)^{op} will be an Artin algebra, so that the category of finitely presented modules coincides with the category of finitely generated modules.

We will be interested in these categories in the case when $\mathcal{a}$ is the category of finitely generated (left) modules over an Artin algebra Λ. We write mod(Λ) = $\mathcal{C}$, and we write $\hat{\mathcal{C}}_0(\Lambda)$ for $\hat{\mathcal{C}}_0$, when necessary. In this case we will be able to describe all projective and injective objects, and show that $\underline{\text{mod}(\Lambda)}$ is equivalent to the full subcategory of projective objects of $\hat{\mathcal{C}}_0(\Lambda)$ and $\overline{\text{mod}(\Lambda)}$ is equivalent to the full subcategory of injective objects. In particular, with our previous notation, $\hat{\mathcal{C}}_0(\Lambda)$ is equivalent to Mod($\underline{\text{mod}(\Lambda)}$) and to Comod($\overline{\text{mod}(\Lambda)}$). This is the setting in which we shall try to obtain more information on stable equivalence.

For the special case that Λ is of finite representation type, $\hat{\mathcal{C}}$ = mod(Γ), where Γ = End(M)^{op}. Here M is the direct sum of one copy of each of the non-isomorphic indecomposable Λ-modules. In this case $\hat{\mathcal{C}}_0(\Lambda)$ = mod($\Gamma/\mathfrak{A}$), where $\mathfrak{A}$ is the ideal of Γ generated by images of maps from (right) injective modules into Γ. Hence what we prove about $\hat{\mathcal{C}}_0(\Lambda)$ in general, will in particular apply to these rings $\Gamma/\mathfrak{A}$.

§3. INJECTIVE AND PROJECTIVE OBJECTS IN $\hat{\mathcal{C}}_0(\Lambda)$. In this section we want to describe the injective and projective objects in $\hat{\mathcal{C}}_0(\Lambda)$.

PROPOSITION 3.1. _If_ G _is half exact and in_ $\hat{\mathcal{C}}_0(\Lambda)$, _then_ G _is injective in_ $\hat{\mathcal{C}}_0(\Lambda)$.

PROOF: Assume that G in $\hat{\mathcal{C}}_0(\Lambda)$ is half exact. We first want to show that for F $\in \hat{\mathcal{C}}_0(\Lambda)$, $\text{Ext}^1_{\hat{\mathcal{C}}}(F,G) = 0$. Since gl. dim. $\hat{\mathcal{C}}(\Lambda) \leq 2$, and the projective objects in $\hat{\mathcal{C}}(\Lambda)$ are the representable functors (,M), we have a projective resolution $0 \to (\ ,A) \to (\ ,B) \to (\ ,C) \to F \to 0$ of F in $\hat{\mathcal{C}}(\Lambda)$.

Since F is in $\hat{\mathcal{C}}_0(\Lambda)$, $F(P) = 0$ for all projective objects in $\mathrm{mod}(\Lambda)$. Since $(\ ,\Lambda)$ is projective in $\hat{\mathcal{C}}(\Lambda)$, the sequence $0 \to ((\ ,\Lambda),(\ ,A)) \to ((\ ,\Lambda),(\ ,B)) \to ((\ ,\Lambda),(\ ,C)) \to ((\ ,\Lambda),F) \to 0$ is exact. By Yoneda, this gives the exact sequence $0 \to A \to B \to C \to 0$ since $F(\Lambda) = 0$. (And conversely, if $0 \to A \to B \to C \to 0$ is exact, then F determined by $0 \to (\ ,A) \to (\ ,B) \to (\ ,C) \to F \to 0$ lies in $\hat{\mathcal{C}}_0(\Lambda)$). Then consider

$$\begin{array}{ccccc} ((\ ,C),G) & \to & ((\ ,B),G) & \to & ((\ ,A),G) \\ \wr\Vert & & \wr\Vert & & \wr\Vert \\ G(C) & & G(B) & & G(A) \end{array}$$

Since G is half exact, the above sequence is exact, which implies that $\mathrm{Ext}^1(F,G) = 0$. If then $0 \to F' \to F \to F'' \to 0$ is an exact sequence in $\hat{\mathcal{C}}_0(\Lambda)$,

$$0 \to (F'',G) \to (F,G) \to (F',G) \to \mathrm{Ext}^1(F'',G)$$

is exact. Since we know that $\mathrm{Ext}^1(F'',G) = 0$, we conclude that G is injective in $\hat{\mathcal{C}}_0(\Lambda)$, which finishes the proof.

The functors $\mathrm{Ext}^1(\ ,A)$ vanish on projectives and are half exact. And the exact sequence

$$(\ ,E(A)) \to (\ ,E(A)/A) \to \mathrm{Ext}^1(\ ,A) \to 0$$

shows that $\mathrm{Ext}^1(\ ,A)$ is coherent. Hence by the above they are injective objects of $\hat{\mathcal{C}}_0(\Lambda)$. We next wnat to show that all injective objects of $\hat{\mathcal{C}}_0(\Lambda)$ are of this type.

PROPOSITION 3.2. <u>The injective objects of</u> $\hat{\mathcal{C}}_0(\Lambda)$ <u>are the functors</u> $\mathrm{Ext}^1(\ ,A)$.

PROOF: Let $0 \to (\ ,A) \to (\ ,B) \to (\ ,C) \to F \to 0$ be a projective resolution of F in $\hat{\mathcal{C}}(\Lambda)$. We have seen that when F is in $\hat{\mathcal{C}}_0(\Lambda)$, $0 \to A \to B \to C \to 0$ is exact. Hence we get the long exact sequence

$$0 \to (\ ,A) \to (\ ,B) \to (\ ,C) \to \mathrm{Ext}^1(\ ,A) \to \mathrm{Ext}^1(\ ,B) \to \mathrm{Ext}^1(\ ,C) \to \dots$$

and consequently an injective resolution of F :

$$0 \to F \to \mathrm{Ext}^1(\ ,A) \to \mathrm{Ext}^1(\ ,B) \to \mathrm{Ext}^1(\ ,C) \to \dots$$

If F is injective, F must be a direct summand of $\mathrm{Ext}^1(\ ,A)$. Since Λ is an Artin algebra, we conclude [3] that F must itself be some $\mathrm{Ext}^1(\ ,A')$.

We go on to describe the projective objects in $\hat{\mathcal{C}}_0(\Lambda)$. Denote by $F = (\ ,\underline{C})$ the functor $F(M) = \mathrm{Hom}_\Lambda(M,C)/R(M,C)$ where $R(M,C)$ is the subgroup of $\mathrm{Hom}_\Lambda(M,C)$ consisting of the maps $f: M \to C$ which factor through a projective Λ-module. $(\ ,\underline{C})$ clearly vanishes on projectives. If P denotes the projective cover of C, then $(\ ,P) \to (\ ,C) \to (\ ,\underline{C}) \to 0$ is exact. For, to say that a map $f: m \to C$ factors through a projective module is equivalent to saying that it factors through the projective cover of C, as is easily seen. We conclude that $(\ ,\underline{C})$ is an object of $\hat{\mathcal{C}}_0(\Lambda)$.

PROPOSITION 3.3. <u>The projective objects of $\hat{\mathcal{C}}_0$ are the functors</u> $(\ ,\underline{C})$.

PROOF: Let P be the projective cover of C, so that $(\ ,P) \to (\ ,C) \overset{\alpha}{\to} (\ ,\underline{C}) \to 0$ is exact. Suppose now that $F \in \hat{\mathcal{C}}_0$. Then $0 \to ((\ ,\underline{C}),F) \overset{\alpha'}{\to} ((\ ,C),F) \to ((\ ,P),F)$ is exact. Since $((\ ,P),F) \cong F(P) = 0$, α' is an isomorphism. If $0 \to F' \to F \to F'' \to 0$ is an exact sequence in $\hat{\mathcal{C}}_0$, then by the above isomorphism and the fact that $(\ ,C)$ is projective in $\hat{\mathcal{C}}$, we get the exact sequence $0 \to ((\ ,\underline{C}),F') \to ((\ ,\underline{C}),F) \to ((\ ,\underline{C}),F'') \to 0$. This shows that $(\ ,\underline{C})$ is projective in $\hat{\mathcal{C}}_0$.

Conversely, let F be a projective object in $\hat{\mathcal{C}}_0$. Consider the diagram

$$\begin{array}{ccccc} & & (\ ,\underline{C}) & & \\ & \nearrow & & & \\ (\ ,C) & \xrightarrow{f} & F & \longrightarrow & 0 . \end{array}$$

By the above isomorphism α', there is a map $g : (\ ,\underline{C}) \to F$ which makes the diagram commute. Since f is epi, we conclude that g is. Since F is assumed to be projective, g splits, so F is a summand of $(\ ,\underline{C})$. Since Λ is an Artin algebra, we conclude that $F = (\ ,\underline{C}')$ [3]. Hence all projectives are of this type.

We now observe that the natural functor from $\text{mod}_P(\Lambda)$, the category of (finitely generated) Λ-modules with no projective summands, to $\hat{\mathcal{C}}_0(\Lambda)$, sending C to $(\ ,\underline{C})$, induces an equivalence of categories between $\underline{\text{mod}(\Lambda)}$ and the full subcategory $\mathcal{P}$ of projective objects in $\hat{\mathcal{C}}_0$.

Similarly, since the morphism $\text{Ext}^1(\ ,A) \to \text{Ext}^1(\ ,B)$ are all of the form $\text{Ext}^1(\ ,f)$ where f is a morphism from A to B and $\text{Ext}^1(\ ,f)$ is zero if and only if f factors through an injective object [8], we get a natural equivalence of categories between $\overline{\text{mod}(\Lambda)}$ and the full subcategory of injective objects of $\hat{\mathcal{C}}_0(\Lambda)$.

We already saw that we had a natural way of constructing an injective resolution for an object F in $\hat{\mathcal{C}}_0$. Namely, if

$$0 \to (\ ,A) \to (\ ,B) \to (\ ,C) \to F \to 0$$

is a projective resolution of F in $\hat{\mathcal{C}}$, then

$$0 \to F \to \text{Ext}^1(\ ,A) \to \text{Ext}^1(\ ,B) \to \cdots$$

gives an injective resolution of F. There is a similar way of constructing a projective resolution for F (see [6]). Consider the exact commutative diagram

$$\begin{array}{ccccccccc}
0 & \to & P_1' & \to & P_1 \oplus P_1' & \to & P_1' & \to & 0 \\
 & & \downarrow & & \downarrow & & \downarrow & & \\
0 & \to & P_0' & \to & P_0 \oplus P_0' & \to & P_0' & \to & 0 \\
 & & \downarrow & & \downarrow & & \downarrow & & \\
0 & \to & A & \to & B & \to & C & \to & 0 \\
 & & \downarrow & & \downarrow & & \downarrow & & \\
 & & 0 & & 0 & & 0 & &
\end{array}$$

where the P_i, P_i' are projective modules. Denote by $\pi_i((\ ,A))$ the ith homology group of the complex

$$\to (\ ,P_i) \to (\ ,P_{i-1}) \to \dots \to (\ ,P_0) \to (\ ,A) \to 0 .$$

We then get a long exact sequence

$$\dots \to \pi_1((\ ,A)) \to \pi_1((\ ,B)) \to \pi_1((\ ,C)) \to \pi_0((\ ,A)) \to \pi_0((\ ,B)) \to$$
$$\pi_0((\ ,C)) \to F \to 0$$

It is easy to see that $\pi_i((\ ,A)) = (\ ,\underline{\Omega^i A})$, where $\Omega^i A$ denotes the ith syzygy module of A. Hence this gives a projective resolution of F

$$\dots \to (\ ,\underline{\Omega^1 C}) \to (\ ,\underline{A}) \to (\ ,\underline{B}) \to (\ ,\underline{C}) \to F \to 0 .$$

In the next section we shall see that $\hat{\mathcal{C}}_0$ has projective covers, and that we can get a minimal projective presentation for F in this way, and also that we can construct a minimal injective copresentation by using the injective resolution mentioned above.

§4. PROJECTIVE PRESENTATIONS AND INJECTIVE COPRESENTATIONS. Before we show the existence of projective covers in $\hat{\mathcal{C}}_0$, we need to show that $\hat{\mathcal{C}}$ has projective covers.

THEOREM 4.1. _An object_ F _in_ $\hat{\mathcal{C}}$ _has a minimal projective cover in_ $\hat{\mathcal{C}}$.

We shall first need the following

LEMMA 4.2. _Let_ $\mathcal{A}$ _be an additive category with the property that all idempotents split,_ M _an object of_ $\mathcal{A}$, _and_ add(M) _the additive category generated by_ M _(i.e. the objects are summands of finite sums of copies of_ M). _If_ B _is in_ $\mathcal{A}$, _and_ $M_0 \in$ add(M), _then the natural map_ $((\ ,M_0), (\ ,B)) \to \mathrm{Hom}_{\mathrm{End}(M)^{op}}((M,M_0), (M,B))$ _is an isomorphism._

PROOF (of the lemma): If $M = M_0$, we are done by Yoneda. To show it for $M_0 = M^n (= M \oplus \dots \oplus M)$, and then for M_0 a summand of M^n, we use additivity.

PROOF (of the theorem): Consider a projective resolution $0 \to (\ ,A) \to (\ ,B) \to (\ ,C) \to F \to 0$ of F. Let $M = B \oplus C$. Then $F(M)$ is a finitely generated $\mathrm{End}(M)^{op}$-module, and $(M,B) \to (M,C) \to F(M) \to 0$ is a projective presentation of $F(M)$ over $\mathrm{End}(M)^{op}$. Let $P_1 \xrightarrow{f'} P_0 \to F(M) \to 0$ be a minimal projective presentation of $F(M)$ over $\mathrm{End}(M)^{op}$, which is an Artin algebra. Since the natural functor $\mathrm{add}(M) \to \mathrm{mod}(\mathrm{End}(M)^{op})$ given by $X \to (M,X)$ is fully faithful, there is a map $f : M_1 \to M_0$ in $\mathrm{add}(M)$ such that if G is defined by $(\ ,M_1) \xrightarrow{f''} (\ ,M_0) \to G \to 0$ being exact, there is a commutative diagram

$$\begin{array}{ccccccc} (M,M_1) & \to & (M,M_0) & \to & G(M) & \to & 0 \\ \wr\| & & \wr\| & & \wr\| & & \\ P_1 & \to & P_0 & \to & F(M) & \to & 0 \end{array}$$

By our lemma the full subcategory of projective objects $(\ ,X)$ of $\hat{\mathcal{C}}$, with X in $\mathrm{add}(M)$, is equivalent to the category of projective $\mathrm{End}(M)^{op}$-modules. Since M_0, M_1, B, C are all in $\mathrm{add}(M)$, and $\mathrm{Coker}((M,M_1) \to (M,M_0))$ and $\mathrm{Coker}((M,B) \to (M,C))$ are isomorphic $(F(M) \cong G(M))$, we conclude that $\mathrm{Coker}((\ ,M_1) \to (\ ,M_0)) = G$ and $\mathrm{Coker}((\ ,B) \to (\ ,C)) = F$ are isomorphic.

We now want to show that $(\ ,M_0) \xrightarrow{f} F \to 0$ is a minimal projective cover. Since $\hat{\mathcal{C}}$ has enough projectives, we need to show that if $(\ ,N) \xrightarrow{h} (\ ,M_0)$ is such that $fh : (\ ,N) \to F$ is an epimorphism, then h is an epimorphism. Consider

$$\begin{array}{c} (\ ,N) \\ \downarrow h \\ (\ ,M_0) \xrightarrow{f} F \to 0 \end{array} \qquad \text{and} \qquad \begin{array}{c} (M,N) \\ \downarrow h' \\ (M,M_0) \xrightarrow{f'} F(M) \to 0 . \end{array}$$

Since (M,M_0) is a projective cover for $F(M)$ in $\mathrm{End}(M)^{op}$, h' is epi, and hence splits. Hence there is a map $g : M_0 \to N \to M_0$ such that $g' : (M,M_0) \to (M,M_0)$

is the identity. Hence g is also the identity, so our original map h splits; hence fh is an epimorphism. This shows that F has a minimal projective cover.

We proceed to show that $\hat{\mathcal{C}}_0$ has minimal projective covers. Let F be an object of $\hat{\mathcal{C}}$. We associate to F an object $\underline{F}$ of $\hat{\mathcal{C}}_0$, defined by the following diagram

$$\begin{array}{ccccccc} (\ ,A) & \to & (\ ,B) & \to & F & \to & 0 \\ \downarrow & & \downarrow & & \downarrow\alpha & & \\ (\ ,\underline{A}) & \to & (\ ,\underline{B}) & \to & \underline{F} & \to & 0 \end{array}$$

where $(\ ,A) \to (\ ,B) \to F \to 0$ is a minimal projective presentation of F. For G in $\hat{\mathcal{C}}_0$ we get the following exact commutative diagram

$$\begin{array}{ccccccc} 0 & \to & (\underline{F},G) & \to & ((\ ,\underline{B}),G) & \to & ((\ ,\underline{A}),G) \\ & & \downarrow & & \downarrow\wr & & \wr\!\!\downarrow \\ 0 & \to & (F,G) & \to & ((\ ,B),G) & \to & ((\ ,A),G) . \end{array}$$

Hence $\alpha : F \to \underline{F}$ induces an isomorphism $\alpha' : (\underline{F},G) \to (F,G)$ for all G in $\hat{\mathcal{C}}_0$. This means that $\alpha : F = \underline{F}$ from $\hat{\mathcal{C}}_0$ to $\hat{\mathcal{C}}$, is a right adjoint to $\mathrm{Id} : \hat{\mathcal{C}}_0 \to \mathcal{C}$.

PROPOSITION 4.3. __If__ $(\ ,C) \to F \to 0$ __is a projective cover in__ $\hat{\mathcal{C}}$, __then__ $(\ ,C) \to \underline{F}$ __is a projective cover in__ $\hat{\mathcal{C}}_0$.

PROOF: Consider the diagram

$$\begin{array}{ccccccc} (\ ,D) & & (\ ,C) & \to & F & \to & 0 \\ \downarrow & & \downarrow & & \downarrow & & \\ (\ ,\underline{D}) & \to & (\ ,\underline{C}) & \to & \underline{F} & \to & 0 \\ \downarrow & & \downarrow & & \downarrow & & \\ 0 & & 0 & & 0 & & \end{array}$$

where the map from $(\ ,\underline{D})$ to $(\ ,\underline{C})$ is such that $(\ ,\underline{D}) \to (\ ,\underline{C}) \to \underline{F}$ is epi. Hence $(\ ,D) \to (\ ,\underline{D}) \to \underline{F}$ is epi, and this map factors as $(\ ,D) \to F \to \underline{F}$.

We shall need the following

LEMMA 4.4. *Let* P *be the projective cover of* C. *Then* $(\ ,P) \to F \to \underline{F} \to 0$ *is exact*.

PROOF: Consider the exact commutative diagram

$$\begin{array}{ccccc}
(\ ,P) & = & (\ ,P) & & \\
\downarrow & & \downarrow & & \\
(\ ,C) & \to & F & \to & 0 \\
\downarrow & & \downarrow & & \\
(\ ,\underline{C}) & \to & G & \to & 0 \\
\downarrow & & \downarrow & & \\
0 & & 0 & &
\end{array}$$

where we want toshow that $G \cong \underline{F}$. We get the exact sequence $0 \to (G,X) \to (F,X) \to ((\ ,P),X)$, i.e. since $((\ ,P),X) = 0$ for X in $\hat{\mathcal{C}}_0$, the isomorphism $(G,X) \xrightarrow{\sim} (F,X) \cong (\underline{F},X)$. We conclude that $G \cong \underline{F}$.

Since $(\ ,D) \to \underline{F}$ is an epimorphism, we now get that $(\ ,D) \oplus (\ ,P) \to F$ is an epimorphism. Since $(\ ,C)$ is a projective cover for F in $\hat{\mathcal{C}}$, $(\ ,D) \oplus (\ ,P) \to (\ ,C)$ is epi. Hence $(\ ,D) \oplus (\ ,P) \to (\ ,\underline{C})$ is epi, and hence also $(\ ,\underline{D}) \oplus (\ ,\underline{P}) \to (\ ,\underline{C})$, i.e. $(\ ,\underline{D}) \to (\ ,\underline{C})$ is epi. This finishes the proof that $(\ ,\underline{C})$ is a projective cover for F.

PROPOSITION 4.5. *Let* $0 \to (\ ,A) \to (\ ,B) \to (\ ,C) \to F \to 0$ *be a minimal projective resolution for* F *in* $\hat{\mathcal{C}}$. *Then* $(\ ,\underline{B}) \to (\ ,\underline{C}) \to F \to 0$ *is a minimal projective presentation for* F *in* $\hat{\mathcal{C}}_0$.

PROOF: Since $F \in \hat{\mathcal{C}}_0$, $0 \to A \to B \to C \to 0$ is exact. Let $P_0 \to A \to 0$, $P_0' \to C \to 0$ be projective covers, and consider the exact commutative diagram

$$\begin{array}{ccccccccc}
0 & \to & P_0 & \to & P_0 \oplus P_0' & \to & P_0' & \to & 0 \\
& & \downarrow & & \downarrow & & \downarrow & & \\
0 & \to & A & \to & B & \to & C & \to & 0 \\
& & \downarrow & & \downarrow & & \downarrow & & \\
& & 0 & & 0 & & 0 & &
\end{array}$$

This gives rise to the exact commutative diagram

$$
\begin{array}{ccccccccccc}
0 & \to & (\ ,P_0) & \to & (\ ,P_0) \oplus (\ ,P_0') & \to & (\ ,P_0') & \to & 0 & & \\
 & & \downarrow & & \downarrow & & \downarrow & & & & \\
0 & \to & (\ ,A) & \to & (\ ,B) & \to & (\ ,C) & \to & F & \to & 0 \\
 & & \downarrow & & \downarrow & & \downarrow & & \| & & \\
 & & (\ ,\underline{A}) & \to & (\ ,\underline{B}) & \to & (\ ,\underline{C}) & \to & F & \to & 0 \\
 & & \downarrow & & \downarrow & & \downarrow & & & & \\
 & & 0 & & 0 & & 0 & & & &
\end{array}
$$

Let H be such that $0 \to H \to (\ ,C) \to F \to 0$ is exact. Then $(\ ,A) \to (\ ,B) \to H \to 0$ is exact, hence by definition of $\underline{H}$, $(\ ,\underline{A}) \to (\ ,\underline{B}) \to \underline{H} \to 0$ is exact. Since $(\ ,\underline{A}) \to (\ ,\underline{B}) \to (\ ,\underline{C}) \to F \to 0$ is exact, we then conclude that $0 \to \underline{H} \to (\ ,\underline{C}) \to F \to 0$ is exact. Since $(\ ,B)$ is a projective cover for H in $\hat{\mathcal{C}}$, $(\ ,\underline{B})$ is a projective cover for $\underline{H}$ in $\hat{\mathcal{C}}_0$ by Proposition 4.3. We conclude $(\ ,\underline{B}) \to (\ ,\underline{C}) \to F \to 0$ is a minimal projective presentation of F.

For injective resolutions we have a corresponding result.

PROPOSITION 4.6. $\mathrm{Ext}^1(\ ,A)$ <u>is an injective envelope for</u> F, <u>and</u> $0 \to F \to \mathrm{Ext}^1(\ ,A) \to \mathrm{Ext}^1(\ ,B)$ <u>is a minimal injective copresentation</u>.

We leave out the direct proof of this result. For we shall now see that there is a natural duality between $\hat{\mathcal{C}}_0(\Lambda)$ and $\hat{\mathcal{C}}_0(\Lambda^{op})$, and using this, it can be deduced from the analogous result on projective resolutions.

The duality $D : \mathrm{mod}(\Lambda) \to \mathrm{mod}(\Lambda^{op})$ given by $D(M) = \mathrm{Hom}_R(M, E_R(R/\mathfrak{m}))$ induces a duality $D : \underline{\mathrm{mod}(\Lambda)} \to \overline{\mathrm{mod}(\Lambda^{op})}$, hence a duality D' from the full subcategory $\mathcal{P}$ of $\hat{\mathcal{C}}_0(\Lambda)$, whose objects are the projective objects of $\hat{\mathcal{C}}_0(\Lambda)$ to the full subcategory $\mathcal{J}$ of $\hat{\mathcal{C}}_0(\Lambda^{op})$, whose objects are the injective objects of $\hat{\mathcal{C}}_0(\Lambda^{op})$. D' is given by $D'(\ ,\underline{A}) = \mathrm{Ext}^1(\ ,D(A))$. Since $\hat{\mathcal{C}}_0(\Lambda) = \mathrm{Mod}(\mathcal{P})$ and $\hat{\mathcal{C}}_0(\Lambda^{op}) = \mathrm{Comod}(\mathcal{J})$, we get induced a duality $D'' : \hat{\mathcal{C}}_0(\Lambda) \to \hat{\mathcal{C}}_0(\Lambda^{op})$.

We point out that our naturally defined injective and projective resolutions are not in general minimal beyond the first two terms.

CHAPTER II

In this chapter we will show how certain properties of Λ-modules C can be expressed as properties of $(\, ,\underline{C})$ or $\mathrm{Ext}^1(\, ,C)$ in $\hat{\mathcal{C}}_0(\Lambda)$. For Λ and Λ' stably equivalent, we let G denote the equivalence $G : \underline{\mathrm{mod}}(\Lambda) \to \underline{\mathrm{mod}}(\Lambda')$, and also the induced map from $\mathrm{mod}_P(\Lambda)$ to $\mathrm{mod}_P(\Lambda')$. Let $V = DT$ be the map from $\mathrm{mod}_P(\Lambda)$ to $\mathrm{mod}_I(\Lambda)$, induced by the natural equivalence $\underline{\mathrm{mod}(\Lambda)} \to \overline{\mathrm{mod}(\Lambda)}$, V^{-1} the inverse map. We let H denote the equivalence $H : \overline{\mathrm{mod}(\Lambda)} \to \overline{\mathrm{mod}(\Lambda')}$, and the induced map $\mathrm{mod}_I(\Lambda) \to \mathrm{mod}_I(\Lambda')$, where $H(M) = VGV^{-1}(M)$. We will interpret our results about $\hat{\mathcal{C}}_0$ to get some necessary conditions on the functors G and H, thus obtaining necessary conditions for two Artin algebras to be stably equivalent.

So we start out by studying $\hat{\mathcal{C}}_0(\Lambda)$. In particular, our previous results are particularly suited for describing when an object F of $\hat{\mathcal{C}}_0(\Lambda)$ has injective or projective dimension 0 or 1, since we know how to constrcut minimal projective presentations and minimal injective copresentations. In particular we will specialize this description to F being $(\, ,\underline{C})$ or $\mathrm{Ext}^1(\, ,A)$.

§1. WHEN IS F PROJECTIVE OR INJECTIVE? Let $0 \to (\, ,A) \to (\, ,B) \to (\, ,C) \to F \to 0$ be a minimal projective resolution of F in $\hat{\mathcal{C}}(\Lambda)$.

PROPOSITION 1.1.

(a) F <u>is projective if and only if</u> B <u>is a projective</u> Λ-<u>module</u>.

(b) F <u>is injective if and only if</u> B <u>is an injective</u> Λ-<u>module</u>.

PROOF: (a) Since $(\, ,\underline{B}) \to (\, ,\underline{C}) \to F \to 0$ is a minimal projective presentation for F, F is projective if and only if $(\, ,\underline{B}) = 0$, which holds if and only if $\underline{B} = 0$. (b) follows similarly by using that $0 \to F \to \mathrm{Ext}^1(\, ,A) \to \mathrm{Ext}^1(\, ,B)$ is a minimal injective copresentation.

An exact sequence $0 \to A \to B \to C \to 0$ of Λ-modules is called the associated minimal sequence of a functor F in $\hat{\mathcal{C}}_0(\Lambda)$ if $0 \to (\ ,A) \to (\ ,B) \to (\ ,C) \to F \to 0$ is a minimal projective presentation for F in $\hat{\mathcal{C}}(\Lambda)$.

We are particularly interested in the cases where $\mathrm{Ext}^1(\ ,A)$ is projective and $(\ ,\underline{C})$ is injective, which we formulate in our next

PROPOSITION 1.2.

(a) $\mathrm{Ext}^1(\ ,A)$ <u>is projective if and only if</u> $E(A)$ <u>is projective, where</u> $E(A)$ <u>is the injective envelope of</u> A.

(b) $(\ ,\underline{C})$ <u>is injective if and only if</u> P_C <u>is injective, where</u> P_C <u>denotes the projective cover of</u> C.

PROOF: (a) If $F = \mathrm{Ext}^1(\ ,A)$, the associated minimal sequence is $0 \to A \xrightarrow{i} E(A) \to E(A)/A \to 0$. This sequence is clearly minimal since i is an essential extension, and

$$0 \to (\ ,A) \to (\ ,E(A)) \to (\ ,E(A)/A) \to \mathrm{Ext}^1(\ ,A) \to 0$$

is exact, since $E(A)$ is injective.

(b) If $F = (\ ,\underline{C})$, the associated minimal sequence is $0 \to \mathrm{Ker}\ f \to P_C \xrightarrow{f} C \to 0$. Again the sequence is clearly minimal, and we have the exact sequence $0 \to (\ ,\mathrm{Ker}\ f) \to (\ ,P_C) \to (\ ,C) \to (\ ,\underline{C}) \to 0$.

COROLLARY 1.3. <u>If</u> Λ <u>and</u> Λ' <u>are stably equivalent, and have no semi-simple ring summands, then</u> Λ <u>has no projective injective module if and only if</u> Λ <u>does not</u>.

PROOF: Let E be an indecomposable projective injective Λ-module. Since Λ has no (semi-) simple ring summand, E cannot be simple. Let $S = \mathrm{soc}\ E$ where $\mathrm{soc}\ E$ denotes the socle of E. Then $\mathrm{Ext}^1(\ ,S)$ is projective by Proposition 1.2, hence $\mathrm{Ext}^1(\ ,H(S))$ is projective in $\hat{\mathcal{C}}_0(\Lambda)$. We remind the reader that H was our map from $\mathrm{mod}_I(\Lambda)$ to $\mathrm{mod}_I(\Lambda')$. This implies that $E(H(S))$ is a projective Λ'-module. This proves the corollary.

We also have the following

COROLLARY 1.4. Let A be an object of $\mathrm{mod}_I(\Lambda)$. $E(A)$ is projective if and only if $E(H(A))$ is projective. In particular, there is a one-one correspondence between the indecomposable non-injective modules of Λ and Λ' such that the injective envelope is projective.

The following dual statement is also true.

COROLLARY 1.5. Let C be an object of $\mathrm{mod}_P(\Lambda)$. P_C is projective if and only if $P_{G(C)}$ is projective, where G was our map from $\mathrm{mod}_P(\Lambda)$ to $\mathrm{mod}_P(\Lambda')$.

PROPOSITION 1.6. $E(\Lambda)$ is projective if and only if $E((\ ,\underline{C}))$ is projective in $\hat{\mathcal{C}}_0(\Lambda)$, for all projectives $(\ ,\underline{C})$.

PROOF: Let C be in $\mathrm{mod}_P(\Lambda)$, and consider the exact sequence

$$0 \to K \to P_C \to C \to 0 .$$

Then

$$0 \to (\ ,\underline{C}) \to \mathrm{Ext}^1(\ ,K) \to \mathrm{Ext}^1(\ ,P_C)$$

is a minimal injective copresentation for $(\ ,\underline{C})$. If $E(\Lambda)$ is projective, certainly $E(K)$ is. Hence by Proposition 1.2, $\mathrm{Ext}^1(\ ,K)$ is projective.

Assume conversely that all $E_0((\ ,\underline{C}))$ are projective where $E_0((\ ,\underline{C}))$ denotes the injective envelope of $(\ ,\underline{C})$, i.e. that $E(K)$ is projective for each K occuring in a minimal sequence $0 \to K \to P_C \to C \to 0$. We want to show that $E(P)$ is projective for all projective indecomposable Λ-modules P. If P is not simple, consider the minimal sequence $0 \to \mathfrak{r}P \to P \to P/\mathfrak{r}P \to 0$. We know that $E(\mathfrak{r}P)$ is projective, and since $\operatorname{soc} P \subset \mathfrak{r}P, E(\mathfrak{r}P) \cong E(P)$. If P is simple projective, it would be sufficient to show that P is contained in some non-simple indecomposable projective module P'. This is taken care of by the following

LEMMA 1.7. <u>Any simple non-injective projective Λ-module is contained in some non-simple indecomposable projective module.</u>

PROOF: Assume that S is a simple non-injective projective module not contained in any other indecomposable projective modules, and denote by P the sum of one copy of each of the other indecomposable projective Λ-modules. Then clearly $\operatorname{Hom}(P,S) = \operatorname{Hom}(S,P) = 0$, hence $\Lambda \cong \operatorname{End}(P)^{op} \times \operatorname{End}(S)^{op}$. This would imply that S is an injective Λ-module, a contradiction.

COROLLARY 1.8. $E(\Lambda)$ <u>is projective if and only if in</u> $0 \to (\ ,\underline{C}) \to E_0((\ ,\underline{C})) \to E_1((\ ,\underline{C}))$, <u>a minimal injective copresentation of</u> $(\ ,\underline{C})$, <u>the injectives are projective for all</u> Λ-<u>modules</u> C.

PROOF: Let $0 \to K \to P_C \to C \to 0$ be a minimal sequence, so that $0 \to (\ ,\underline{C}) \to \operatorname{Ext}^1(\ ,K) \to \operatorname{Ext}^1(\ ,P_C)$ is a minimal injective copresentation for $(\ ,\underline{C})$. Since $E(\Lambda)$ projective implies $\operatorname{Ext}^1(\ ,P_C)$ projective, our corollary follows from Proposition 1.6.

COROLLARY 1.9. <u>If</u> Λ <u>and</u> Λ' <u>are stably equivalent, then</u> Λ <u>is 1-Gorenstein (i.e.</u> $E(\Lambda)$ <u>is projective) if and only if</u> Λ' <u>is</u>.

PROOF: An immediate consequence of Corollary 1.8.

We remark that we can also give a description of when gl. dim. $\hat{\mathcal{C}}_0(\Lambda) = 0$, but we leave out the proof of this result here.

PROPOSITION 1.10. gl. dim $\hat{\mathcal{C}}_0(\Lambda) = 0$ <u>if and only if</u> $\mathfrak{r}^2 = 0$ <u>and</u> Λ <u>is 1-Gorenstein</u>.

§2. WHEN DOES F HAVE PROJECTIVE (INJECTIVE) DIMENSION 1? Again we let $0 \to A \xrightarrow{f} B \xrightarrow{g} C \to 0$ denote the minimal sequence associated with F.

PROPOSITION 2.1.

(a) $\operatorname{pd} F \leq 1$ *if and only if* $f : A \to B$ *factors through a projective module*.

(b) $\operatorname{id} F \leq 1$ *if and only if* $g : B \to C$ *factors through an injective module*.

PROOF: Since $(\ ,\underline{A}) \xrightarrow{f'} (\ ,\underline{B}) \to (\ ,\underline{C}) \to F \to 0$ gives a projective resolution for F, minimal in the first two terms, then $\operatorname{pd} F \leq 1$ if and only if $f' : (\ ,\underline{A}) \to (\ ,\underline{B})$ is zero. This is equivalent to saying that $f : A \to B$ factors through a projective module.

$0 \to F \to \operatorname{Ext}^1(\ ,A) \to \operatorname{Ext}^1(\ ,B) \xrightarrow{g'} \operatorname{Ext}^1(\ ,C)$ is an injective resolution of F which is minimal in the first two terms. Hence $\operatorname{id} F \leq 1$ if and only if $g' : \operatorname{Ext}^1(\ ,B) \to \operatorname{Ext}^1(\ ,C)$ is zero, which is the case if and only if $g : B \to C$ factors through an injective module.

We specialize again to $F = (\ ,\underline{C})$ and $F = \operatorname{Ext}^1(\ ,A)$. We say that a module A is torsionless if A is a submodule of a (finitely generated) projective module, cotorsionless if it is a factor module of an injective module.

PROPOSITION 2.2. *Let* A *be an indecomposable non-injective* Λ*-module*, C *an indecomposable non-projective* Λ*-module. Then*

(a) $\operatorname{pd} \operatorname{Ext}^1(\ ,A) \leq 1$ *if and only if* A *is torsionless*

(b) $\operatorname{id}(\ ,\underline{C}) \leq 1$ *if and only if* C *is cotorsionless*.

PROOF: Since $0 \to A \to E(A) \to E(A)/A \to 0$ is the minimal sequence corresponding to $\operatorname{Ext}^1(\ ,A)$, we need to show that A is torsionless if and only if $A \xrightarrow{i} E(A)$ factors through a projective module. If we have a commutative diagram

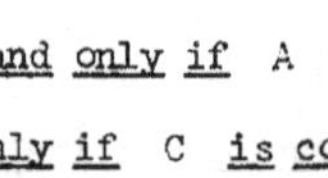

with P projective, then j must be a monomorphism. And if A is a submodule of some projective module P, then $i : A \to E(A)$ can be extended to a map $k : P \to E(A)$, since $E(A)$ is injective. Hence (a) follows.

(b) follows similarly by considering the exact sequence $0 \to K \to P_C \xrightarrow{g} C \to 0$, and observing that $g : P_C \to C$ factors through an injective module if and only if C is a factor module of an injective module.

COROLLARY 2.3.

(a) H <u>gives a one-one correspondence between the indecomposable non-injective torsionless modules of</u> Λ <u>and</u> Λ'.

(b) G <u>gives a one-one correspondence between the indecompsable non-projective cotorsionless modules</u>.

§3. THE INJECTIVE COPRESENTATION OF $(,\underline{C})$. In this section we investigate for which indecomposable non-injective Λ-modules M, $\mathrm{Ext}^1(,M)$ occurs as a summand in the $E_0((,\underline{C}))$ or in the $E_1((,\underline{C}))$ for some projective object $(,\underline{C})$ in $\hat{\mathcal{C}}_0$.

We recall that a minimal injective copresentation for $(,\underline{C})$ is given by $0 \to (,\underline{C}) \to \mathrm{Ext}^1(,K) \to \mathrm{Ext}^1(,P_C)$, where $0 \to K \to P_C \to C \to 0$ is exact.

PROPOSITION 3.1. $\mathrm{Ext}^1(,M)$, <u>where</u> M <u>is an indecomposable non-injective</u> Λ-<u>module is a summand of some</u> $E_1((,C))$ <u>if and only if</u> M <u>is a non-simple projective</u> Λ-<u>module</u>.

PROOF: Let $M \doteq P$ be a non-simple indecomposable projective Λ-module, and consider $0 \to \mathfrak{R}P \to P \to S \to 0$. Then $\mathrm{Ext}^1(,P) = E_1((,\underline{S}))$.

Since all $E_1((,\underline{C}))$ are of the type $\mathrm{Ext}^1(,P_C)$, where P_C is some projective cover of C, it is clear that each $\mathrm{Ext}^1(,M)$ which is a summand of some $E_1((,\underline{C}))$, has M projective. But if C is an indecomposable non-projective Λ-module, then no summands of P_C can be simple. For this would imply that this simple projective module would also be a summand of C, a contradiction. This finishes the proof of the proposition.

COROLLARY 3.2. $H : mod_I(\Lambda) \to mod_I(\Lambda')$ *gives a one-one correspondence between the indecomposable non-simple non-injective projective modules over* Λ *and* Λ'. *In particular, the number is the same.*

PROPOSITION 3.3. $Ext^1(\ ,M)$, *for* M *an indecomposable non-injective* Λ-*module, occurs as a summand in some* $E_0((\ ,\underline{C}))$ *if and only if* M *is a torsionless module contained in* $\mathfrak{R}P$ *for some projective module* P.

PROOF: This is immediate, since the summands of $E_0((\ ,\underline{C}))$ are given by $Ext^1(\ ,K)$, where K comes from a minimal sequence $0 \to K \to P_C \to C \to 0$; i.e. from a sequence where $K \subset \mathfrak{R}P_C$.

COROLLARY 3.4. *If* Λ *and* Λ' *are stably equivalent, and* $\mathfrak{R}^2 = \mathfrak{R}'^2 = 0$, *then* H *gives a one-one correspondence between the non-injective simple modules.*

PROOF: For $\mathfrak{R}^2 = 0$, the indecomposable non-injective modules M which are contained in $\mathfrak{R}P$ for some projective module are exactly the simple torsionless modules. We further need to show that each non-injective simple module is torsionless in this case: Let S be simple and not injective, and let P be the projective cover of $E(S)$. Then $\mathfrak{R}p$ maps onto $\mathfrak{R}E(S) = S$, hence S is contained in P since $\mathfrak{R}P$ and $\mathfrak{R}E(S)$ are semi-simple modules.

COROLLARY 3.5. *If* Λ *and* Λ' *are stably equivalent and have no semi-simple ring summands and* $\mathfrak{R}^2 = \mathfrak{R}'^2 = 0$, *then* Λ *and* Λ' *have the same number of indecomposable projective injective modules.*

PROOF: We have seen that Λ and Λ' have the same number of non-injective simple modules. By duality they have the same number of non-simple projective modules. But Λ and Λ' also have the same number of indecomposable non-simple non-injective projective modules, hence the same number of non-simple projective injective modules. Since the rings have no semi-simple ring summands, no module is simple, projective and injective. This proves our corollary.

We will also point out what we can conclude about the top of C (i.e. $C/\mathfrak{R}C$) from a minimal injective copresentation $0 \to (\ ,\underline{C}) \to \mathrm{Ext}^1(\ ,K) \to \mathrm{Ext}^1(\ ,P)$, of $(\ ,\underline{C})$ where $0 \to K \to P \to C \to 0$ is exact. Let $C/\mathfrak{R}C = S_1 \oplus \dots \oplus S_n \oplus T_1 \oplus \dots \oplus T_m$, where the S_i are simples with P_{S_i} not injective, the T_j simples with P_{T_j} injective. We then see that the number n is the number of indecomposable summands of $E_1((\ ,\underline{C}))$, hence n is also the number of simples in the top of $G(C)$ whose projective covers are not injective. The map g between the non-projective simple modules whose projective cover is not injective, inducing the maps from part of $C/\mathfrak{R}C$ to part of $G(C)/\mathfrak{R}'G(C)$, is then given as follows: $g(S) =$ the top of $H(P_S)$.

If Λ, and hence Λ', has no projective injective modules, we get a closer connection:

PROPOSITION 3.6. *If Λ and Λ' are stably equivalent, and there are no projective injective modules, then there is a one-one correspondence g between the simple non-projective modules, such that g carries the top of C to the top of $G(C)$. In particular $\ell(C/\mathfrak{R}C) = \ell(G(C)/\mathfrak{R}'G(C))$, where $\ell(X)$ is the length of the module X.*

PROOF: This follows from our discussion above, since there are no simple modules with injective projective covers. We saw that g was given by $g(S) = H(P_S)/\mathfrak{R}'H(P_S)$. But choosing in particular $C = S$, we must have $g(S) = G(S)$.

Similarly, if A is an indecomposable non-injective Λ-module, we consider $0 \to A \to E(A) \to E(A)/A \to 0$, and $(\ ,E(A)) \to (\ ,E(A)/A) \to \mathrm{Ext}^1(\ ,A) \to 0$. We now get that there is a correspondence between the simples whose injective envelopes are not projective, in soc A and soc H(A). In particular, we get

PROPOSITION 3.7. *If* Λ *and* Λ' *have no projective injective modules,* H *gives a correspondence between the non-injective simple modules over* Λ *and* Λ', *which carries* soc A *to* soc H(A). *In particular,* $\ell(\text{soc } A) = \ell(\text{soc } H(A))$.

Using the results of this section, we can give a description of the Artin algebras Λ which have the property that gl. dim $\hat{\mathcal{C}}_0(\Lambda) \leq 1$. We do not provide the proof of this fact here.

THEOREM 3.8. gl. dim. $\hat{\mathcal{C}}_0(\Lambda) \leq 1$ *if and only if* $\mathfrak{r}^2 = 0$, *and each indecomposable* Λ*-module is simple, projective or injective.*

For an hereditary algebra Λ, we have gl. dim $\hat{\mathcal{C}}_0(\Lambda) \leq 2$, hence this is the case for all Λ' stably equivalent to an hereditary algebra. More generally, we have

PROPOSITION 3.9. *If* gl. dim. $\Lambda \leq n$, *then* gl. dim $\hat{\mathcal{C}}_0(\Lambda) \leq 3n - 1$.

PROOF: Let F be an object of $\hat{\mathcal{C}}_0(\Lambda)$, and let $0 \to A \to B \to C \to 0$ be the associated minimal sequence. Then

$$0 \to F \to \text{Ext}^1(\ ,A) \to \text{Ext}^1(\ ,B) \to \text{Ext}^1(\ ,C) \to \text{Ext}^2(\ ,A) \to \cdots$$

is an injective resolution for F, from which the result follows.

CHAPTER III

In this chapter we investigate when two hereditary algebras are stably equivalent. To handle this problem, we first prove a structure theorem for hereditary algebras, namely that each hereditary algebra Λ can be decomposed into a product of two algebras, where one has the property that the injective envelope of the ring is projective, and the other one has no projective injectives. Assuming as usual that $\Lambda/\mathfrak{R}$ is a product of division rings, we show that the indecomposable hereditary algebras of the first type are $Tr_{n_i}(D_i)$ = the full lower triangular $n_i \times n_i$ matrix ring with entries in a division ring D_i. We then go on to show that up to (Morita equivalence and) a semi-simple part, there is only one hereditary algebra in a stable equivalence class. In the next chapter we will characterize the Artin algebras which are stably equivalent to some hereditary algebra.

§1. STRUCTURE OF HEREDITARY ALGEBRAS.

PROPOSITION 1.1. *A hereditary algebra* Λ *(with no semi-simple ring summand) is the product of two hereditary algebras* Λ_1 *and* Λ_2, *where* Λ_1 *is 1-Gorenstein (i.e.* $E_{\Lambda_1}(\Lambda_1)$ *is projective) and* Λ_2 *has the property that no projective* Λ_2*-module is injective.*

PROOF: Let $P_1, \ldots, P_r$ denote the indecomposable projective Λ-modules with $E(P_i)$ projective, $Q_1, \ldots, Q_s$ the indecomposable projective Λ-modules with $E(Q_j)$ not projective. Let $P = P_1 \oplus \ldots \oplus P_r$, $Q = Q_1 \oplus \ldots \oplus Q_r$. We will need the following

LEMMA 1.2. *If* Λ *is an hereditary algebra, and* M *is an indecomposable* Λ*-module with a simple module* $S \subset M$, *such that* $E(S)$ *is projective, then* $M \subset E(S)$; *in particular* $E(M)$ *is projective.*

PROOF (of the lemma): Consider the diagram

$$\begin{array}{ccccc} 0 & \to & S & \xrightarrow{j} & M \\ & & i\downarrow & & \\ & & E(S) & & \end{array}$$

and let $f: M \to E(S)$ be a map which makes it commute. Since $E(S)$ is projective and Λ is hereditary, Im f is projective. Because M is indecomposable, we conclude that $M \cong \operatorname{Im} f \subset E(S)$. This finishes the proof of the lemma.

Since Λ is hereditary, any non-zero map between indecomposable projective modules must be an injection. If $Q_j \subset P_i$, then $E(Q_j) \subset E(P_i)$. Hence $E(Q_j)$ would be a summand of $E(P_i)$, and consequently projective, a contradiction. Hence $\operatorname{Hom}(Q_j, P_i) = 0$.

If $P_i \subset Q_j$, let S be a simple module with $S \subset P_i \subset Q_j$. By assumption $E(S) \subset E(P_i)$ is projective, so by Lemma 1.2 $Q_j \subset E(S)$, a contradiction to $E(Q_j)$ being not projective. Hence $\operatorname{Hom}(P_i, Q_j) = 0$.

We now get that $\operatorname{End}_\Lambda(\Lambda)^{op} \cong \operatorname{End}(P \oplus Q)^{op} \cong (\operatorname{End}(P) \times \operatorname{End}(Q))^{op} \cong \operatorname{End}(P)^{op} \times \operatorname{End}(Q)^{op} = \Lambda_1 \times \Lambda_2$. The P_i are now the projective indecomposable Λ_1-modules, and $E_{\Lambda_1}(P_i)$ is injective, since $E_\Lambda(P_i)$ is. And the Q_j are the indecomposable projective Λ_2-modules, and $E_{\Lambda_2}(Q_j)$ is not projective. If Λ_2 had some indecomposable projective injective module, it must be some Q_j. But $E_{\Lambda_2}(Q_j) \neq Q_j$, since $E_{\Lambda_2}(Q_j)$ is not projective.

PROPOSITION 1.3. <u>An hereditary 1-Gorenstein algebra</u> Λ <u>is a finite product of rings of type</u> $Tr_{n_i}(D_i)$, <u>a lower</u> $n_i \times n_i$ <u>triangular matrix ring with entries in a division ring</u> D_i.

PROOF: Let P be a projective indecomposable Λ-module. If S is a simple submodule of P, then since E(S) is projective, Lemma 1.2 gives $P \subset E(S)$, hence P has simple socle. Any projective injective module E(S) is uniserial, since all $\mathfrak{r}^i E(S)$ are indecomposable, because they have simple socel, and are projective, since Λ is hereditary. Hence for a projective simple module S, we have a chain of the indecomposable projective modules with socle S:

$$S = P_1 \subset P_2 \subset \dots \subset P_n = E(S).$$

If Q is an indecomposable projective module with $\operatorname{soc} Q = T \neq S$, then $\operatorname{Hom}(P_i,Q) = 0 = \operatorname{Hom}(Q,P_i)$. For an inclusion map is impossible since the socles are different.

Letting $S_1, \ldots, S_r$ denote the simple projective Λ-modules, and Q_i the sum of the indecomposable projective modules with socle S_i, then

$$\Lambda = \operatorname{End}(\Lambda)^{op} \cong \operatorname{End}(Q_1)^{op} \times \cdots \times \operatorname{End}(Q_r)^{op} .$$

Hence to finish the proof of our proposition it is sufficient to show that $\operatorname{End}(P_1 \oplus \cdots \oplus P_n)^{op}$ is $Tr_n(D)$, where $D = \operatorname{End}_\Lambda(S)^{op}$.

We first show that $\operatorname{End}(P_i)^{op} = D$: Let $f : P_i \to P_i$ be a non-zero map, hence an inclusion. Then f induces an isomorphism $f' : S \to S$. And if conversely we start with an isomorphism $f' : S \to S$, then since $E(S)$ is injective, there is a map $f'' : E(S) \to E(S)$, extending $f' : S \to S$. Since $P_i = \mathfrak{r}^{n-i}E(S)$, $f''(P_i) = \mathfrak{r}^{n-i}f(E(S)) \subset \mathfrak{r}^{n-i}E(S) = P_i$, so we get induced a map $f : P_i \to P_i$. If $h : P_i \to P_i$ was a map extending $f' : S \to S$; then $(f - h)(S) = 0$, hence $f - h$ is not a monomorphism, and consequently $f - h = 0$. This shows that $\operatorname{End}(P_i)^{op} \cong \operatorname{End}(S)^{op} = D$.

If $P_i \not\subset P_j$, then $\operatorname{Hom}(P_i,P_j) = 0$, since the only possible non-zero maps are inclusion maps. If $P_i \subset P_j$, we want to show that $\operatorname{Hom}(P_i,P_j) \cong D$ as a two-sided D-module. As above we get that any non-zero map $f : P_i \to P_j$ determines a non-zero map $f' : S \to S$ by restriction. And also conversely does a non-zero map $f' : S \to S$ extend uniquely to a non-zero map $f : P_i \to P_j$. This finishes the proof of the proposition.

§2. WHEN ARE TWO HEREDITARY ALGEBRAS STABLY EQUIVALENT? In this section we find conditions for two hereditary algebras to have equivalent stable categories.

We first point out that if Λ is hereditary, then $\underline{\operatorname{mod}(\Lambda)}$ is equivalent to $\operatorname{mod}_P(\Lambda)$. For if M and N are Λ-modules, $f : M \to N$ a non-zero map and P a projective Λ-module such that the diagram

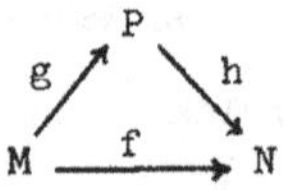

commutes, then $\operatorname{Im} g \neq 0$ is projective, and hence isomorphic to a summand of M. And similarly $\overline{\operatorname{mod}(\Lambda)}$ is equivalent to $\operatorname{mod}_I(\Lambda)$.

THEOREM 2.1. *If Λ and Λ' are hereditary algebras, then Λ and Λ' are stably equivalent if and only if there are semi-simple algebras S and S' such that $\operatorname{mod}(\Lambda \times S)$ and $\operatorname{mod}(\Lambda' \times S')$ are equivalent. (i.e. $\Lambda \times S$ and $\Lambda' \times S'$ isomorphic, with our choice of algebras).*

PROOF: We have already remarked in the introduction that if S is a semi-simple algebra, it is easy to see that Λ and $\Lambda \times S$ are stably equivalent. Hence we assume that Λ and Λ' have no semi-simple ring summands, and are stably equivalent. We then want to show that $\Lambda \cong \Lambda'$.

One of the main problems in trying to get information about the original algebra Λ from $\overline{\operatorname{mod}(\Lambda)}$ is that there is in general no way of recognizing a projective Λ-module when regarded as an object of the category $\overline{\operatorname{mod}(\Lambda)}$. We have seen, however, in Chapter II, that there is a way of recognizing the indecomposable non-injective torsionless modules. This is what enables us to handle the hereditary case. So an important part of the proof of our theorem is provided by the next

LEMMA 2.2. *Let Λ and Λ' be stably equivalent hereditary Artin algebras, with no projective injective modules. Then $\Lambda \cong \Lambda'$.*

PROOF: Let $H : \operatorname{mod}_I(\Lambda) \to \operatorname{mod}_I(\Lambda')$ be the given equivalence of categories. We have seen in Chapter II that H gives a one-one correspondence between the indecomposable non-injective torsionless modules for Λ and Λ'. Since Λ and Λ' are hereditary and no projective modules are injective, this specializes to a correspondence between the indecomposable projective modules of the two rings. Let $P_1, \ldots, P_n$ be the indecomposable projective Λ-modules. Then $H(P_1), \ldots, H(P_n)$ are the indecomposable projective Λ'-modules. We have

$$\Lambda \cong \operatorname{End}_{\Lambda}(\Lambda)^{op} \cong \operatorname{End}_{\Lambda}(P_1 \oplus \cdots \oplus P_n)^{op} \cong \operatorname{End}_{\Lambda'}(H(P_1) \oplus \cdots \oplus H(P_n))^{op} \cong \Lambda' .$$

We point out that we are using the fact that $\overline{\mathrm{mod}_I(\Lambda)}$ is a full subcategory of $\mathrm{mod}(\Lambda)$ when Λ is hereditary.

We want to show a corresponding result for 1-Gorenstein hereditary algebras Λ. That no hereditary algebra with no projective injective modules could be stably equivalent to a 1-Gorenstein hereditary algebra is also taken care of by our next

LEMMA 2.3. Let Λ be an indecomposable hereditary algebra. Then Λ is 1-Gorenstein if and only if $\underline{\mathrm{mod}}(\Lambda)$ is abelian. If Λ is 1-Gorenstein, then $\underline{\mathrm{mod}}(\Lambda)$ and $\mathrm{mod}(\Lambda/\operatorname{soc}\Lambda)$ are equivalent.

PROOF: We remark that since $\underline{\mathrm{mod}}(\Lambda)$ and $\overline{\mathrm{mod}}(\Lambda)$ are equivalent, it does not matter which one we consider. Assume first that $\underline{\mathrm{mod}}(\Lambda)$ is abelian. Assume that Λ is not 1-Gorenstein. Then there is an indecomposable projective Λ-module P with $E(P)$ not projective. If S is a simple module contained in P, with $E(S)$ projective, then by Lemma 1.2, $E(P)$ is projective. So we conclude that $E(P)$ has no projective summands. Clearly $E(P) \xrightarrow{f} E(P)/P$ is a map in $\underline{\mathrm{mod}}(\Lambda)$, which is an epimorphism. We now show that f is also a monomorphism: For given a map $g : X \to E(P)$ such that $fg = 0$, we have $g(X) \subset P$. But if $g \neq 0$, then $g(X)$ is a non-zero projective module which is a summand of X, a contradiction. Hence we conclude that $g = 0$, which shows that f is a monomorphism. Clearly f is not an isomorphism. Hence $\underline{\mathrm{mod}}(\Lambda)$ is not abelian since in an abelian category a morphism which is both a monomorphism and an epimorphism is an isomorphism. So our assumption that $E(P)$

was not projective was wrong, and we conclude that Λ is 1-Gorenstein.

Assume now conversely that Λ is 1-Gorenstein. If $M^* = 0$, where $M^* = \mathrm{Hom}_\Lambda(M, \Lambda)$, then $\mathrm{soc}\,\Lambda \cdot M = 0$. For if $a \cdot m \neq 0$ for some $a \in S$, a simple module contained in $\mathrm{soc}\,\Lambda$, then M would contain a module isomorphic to S, which is projective, since Λ is hereditary. But then we would get $E(S)$ as a summand of $E(M)$, which is impossible since $E(S)^* \neq 0$, $E(M)^* = 0$.

If $M^* \neq 0$, we know that M contains an indecomposable simple projective Λ-module $P = \Lambda e$, for some primitive indempotent e. Now $e \in \mathrm{soc}\,\Lambda$, and $e \cdot e = e \neq 0$, i.e. $\mathrm{soc}\,\Lambda \cdot M \neq 0$.

We have then shown that for Λ 1-Gorenstein, $M^* = 0$ if and only if M is annihilated by $\mathrm{soc}\,\Lambda$. Hence we conclude that $\underline{\mathrm{mod}(\Lambda)}$ and $\mathrm{mod}(\Lambda/\mathrm{soc}\,\Lambda)$ are equivalent in this case. Since $\mathrm{mod}(\Lambda/\mathrm{soc}\,\Lambda)$ is abelian, it follows that $\underline{\mathrm{mod}(\Lambda)}$ is abelian, if Λ is 1-Gorenstein. This finishes the proof of the lemma.

In particular, it follows that $\underline{\mathrm{mod}\,\mathrm{Tr}_n(D)}$ is equivalent to $\mathrm{mod}\,\mathrm{Tr}_{n-1}(D)$. Hence we see that if two 1-Gorenstein hereditary algebras Λ and Λ' are stably equivalent, then $\mathrm{mod}(\Lambda)$ and $\mathrm{mod}(\Lambda')$ are equivalent.

If $\mathcal{A}$ and $\mathcal{B}$ are additive categories, their product $\mathcal{A} \times \mathcal{B}$ is the additive category where the objects are pairs (A,B), where A is an object of $\mathcal{A}$ and B an object of $\mathcal{B}$ and the maps $(A,B) \to (A',B')$ are pairs of maps (f,g), $f \in \mathrm{Hom}(A,A')$, $g \in \mathrm{Hom}(B,B')$. $\mathcal{A}$ is an indecomposable additive category if it cannot be written in the form $\mathcal{A} = \mathcal{A}_1 \times \mathcal{A}_2$ with non-trivial additive categories $\mathcal{A}_1$ and $\mathcal{A}_2$. In particular, Λ is indecomposable if and only if the category $\mathrm{mod}(\Lambda)$ is indecomposable.

We leave out the proof for the following

LEMMA 2.4. *Let* $F : \mathcal{A} \to \mathcal{B}$ *be an equivalence of the additive categories* $\mathcal{A}$ *and* $\mathcal{B}$. *Let* $\mathcal{A} \cong \mathcal{A}_1 \times \cdots \times \mathcal{A}_r$ *and* $\mathcal{B} \cong \mathcal{B}_1 \times \cdots \times \mathcal{B}_s$ *be decompositions of* $\mathcal{A}$ *and* $\mathcal{B}$ *into products of indecomposable categories. Then* $r = s$, *and* F

<u>induces equivalences</u> $F_i : \mathcal{J}_i \to \mathcal{R}_i$ <u>(after change of the numbering)</u>.

If Λ has no semi-simple ring summand and $\overline{\mathrm{mod}(\Lambda)}$ is indecomposable, it is easy to see that Λ is indecomposable. However, it is not in general true that Λ indecomposable implies $\overline{\mathrm{mod}(\Lambda)}$ indecomposable. But we shall need that it be the case if Λ is hereditary.

LEMMA 2.5. <u>For an indecomposable hereditary algebra</u> Λ, $\overline{\mathrm{mod}(\Lambda)}$ <u>is also indecomposable</u>.

PROOF: Assume Λ is indecomposable. Then we have two cases to consider. If no projective Λ-module is injective, then all indecomposable projective Λ-modules are in $\mathrm{mod}_I(\Lambda)$. If $\mathrm{mod}_I(\Lambda)$ decomposes, all the projective Λ-modules must then belong to the same component, since Λ is indecomposable. But since for each M in $\mathrm{mod}_I(\Lambda)$, there is a non-zero map from some projective module, we conclude that $\overline{\mathrm{mod}(\Lambda)}$ is indecomposable.

If $\Lambda = \mathrm{Tr}_n(D)$, then $\mathrm{mod}_I(\Lambda)$ is equivalent to $\mathrm{mod}(\Lambda/\mathrm{soc}\ \Lambda) = \mathrm{mod}\ \mathrm{Tr}_{n-1}(D)$, which is also indecomposable.

We are now ready to put the lemmas together to obtain a proof of the theorem. Decompose Λ as $\Lambda = (\Omega_1 \times \cdots \times \Omega_r) \times (\Gamma_1 \times \cdots \times \Gamma_s)$ where the Ω_i are indecomposable 1-Gorenstein algebras, the Γ_j are indecomposable algebras with no projective module injective. Then $\underline{\mathrm{mod}(\Lambda)} = \underline{\mathrm{mod}(\Omega_1)} \times \cdots \times \underline{\mathrm{mod}(\Gamma_s)}$, a product of indecomposable categories by Lemma 2.5. Assume now that $\underline{\mathrm{mod}(\Lambda)}$ and $\underline{\mathrm{mod}(\Lambda')}$ are equivalent. We have a corresponding decomposition for $\underline{\mathrm{mod}(\Lambda')}$, and by Lemma 2.4 we get induced equivalences between the indecomposable components. By Lemma 2.3 the $\underline{\mathrm{mod}(\Omega_i)}$ are abelian, the $\underline{\mathrm{mod}(\Gamma_j)}$ not. Hence we get the $\underline{\mathrm{mod}(\Omega_i)}$ and $\underline{\mathrm{mod}(\Omega_i')}$ corresponding to each other, and the $\underline{\mathrm{mod}(\Gamma_j)}$ and $\underline{\mathrm{mod}(\Gamma_j')}$. Hence we are reduced to the case of considering indecomposable algebras of the two types, and our theorem follows by Lemma 2.2 and Lemma 2.3.

We describe a map from hereditary algebras with no semi-simple ring summands to hereditary algebras with possibly a semi-simple ring summand, needed in the next chapter. Let Λ be an hereditary algebra. Define $\alpha(\Lambda)$ to be the algebra $\text{End}_{\Lambda}(P)^{op}$, where P is the sum of one copy of each of the indecomposable projective non-injective Λ-modules. By our results, we can give a more explicit description of $\alpha(\Lambda)$. Let $\Lambda = \Lambda_1 \times \prod_{i=1}^{n} \text{Tr}_{m_i}(D_i)$, where Λ_1 has the property that no projective module is injective. $\alpha(\Lambda_1) = \Lambda_1$, and $\alpha(\text{Tr}_{m_i}(D_i)) = \text{Tr}_{m_{i-1}}(D_i)$. Hence $\alpha(\Lambda) = \Lambda_1 \times \prod_{i=1}^{n} \text{Tr}_{m_{i-1}}(D_i)$.

We can also define the inverse α^{-1}. This will be needed in the next chapter.

CHAPTER IV

In this chapter we want to characterize the Artin algebras which are stably equivalent to hereditary algebras. We first show some connection between the lengths of modules over two stably equivalent algebras Λ and Λ', in the case that $\overline{\text{mod}(\Lambda)}$ has kernels. In particular, this is the case if one of the algebras is hereditary. Using these properties, along with results established in previous chapters, we deduce some necessary conditions for an Artin algebra to be stably equivalent to an hereditary algebra. We finally show that two of the conditions are also sufficient, hence providing a description of this class of algebras. These conditions are:

(1) Each indecomposable submodule of an indecomposable projective module is simple or projective.

(2) If S is a simple torsionless non-projective Λ-module, then $S \subset E/\mathfrak{r}E$ for some injective Λ-module E.

We show that these conditions are easily satisfied for algebras with $\mathfrak{r}^2 = 0$, so we get a big class of examples of such rings. In the last section we provide some more examples, for $\mathfrak{r}^2 \neq 0$. We also give some results on this class of rings which in particular show how to construct new examples from old ones. Applying this process to algebras with $\mathfrak{r}^2 = 0$ will give several examples where $\mathfrak{r}^2 \neq 0$. And we will also see that the second (or the first) condition of our theorem cannot be left out.

§1. NECESSARY CONDITIONS. We will first show that if $\overline{\text{mod}(\Lambda)}$ has kernels, we can compare the length of a Λ-module M from $\text{mod}_I(\Lambda)$ with the length of $\overline{M}$ in $\overline{\text{mod}(\Lambda)}$. By the length of $\overline{M}$, $\ell(\overline{M})$, we mean the length of the longest possible chain of monomorphisms (which are not isomorphisms). We start with the following general lemma for arbitrary Artin algebras.

LEMMA 1.1. <u>Let</u> $f : M \to N$ <u>be a morphism in</u> $\text{mod}_I(\Lambda)$. <u>Then</u>:

(a) <u>If</u> $f : M \to N$ <u>is a non-zero monomorphism, then</u> $\overline{f} : \overline{M} \to \overline{N}$ <u>is not zero</u>.

(b) <u>If</u> $\overline{f} : \overline{M} \to \overline{N}$ <u>is a monomorphism in</u> $\overline{\text{mod}(\Lambda)}$, <u>then</u> $f : M \to N$ <u>is a monomorphism in</u> $\text{mod}_I(\Lambda)$.

PROOF: Assume that $f : M \to N$ is a non-zero monomorphism and $\overline{f} : \overline{M} \to \overline{N}$ is zero. Then we have a commutative diagram

$$\begin{array}{ccccc} 0 & \longrightarrow & M & \xrightarrow{f} & N \\ & & \downarrow{\scriptstyle i} & \nearrow{\scriptstyle g} & \\ & & E(M) & & \end{array}$$

where $E(M)$ is the injective envelope of M. Since f is a monomorphism and $i : M \to E(M)$ is an essential extension, g is also a monomorphism. Hence $g : E(M) \to N$ splits, since $E(M)$ is injective, so that $E(M)$ is a summand of N, a contradiction, since N is in $\text{mod}_I(\Lambda)$. We conclude that $\overline{f}$ is not zero.

Assume now that $\overline{f} : \overline{M} \to \overline{N}$ is a monomorphism. Let $L = \text{Ker } f$, so that $0 \to L \xrightarrow{h} M \xrightarrow{f} N \to 0$ is exact. Since $fh = 0$, $\overline{fh} = \overline{f}\,\overline{h} = 0$. Since $\overline{f}$ is a monomorphism, $\overline{h} : \overline{L} \to \overline{M}$ must be zero. By (a) we then conclude that $h : L \to M$ is zero, since h is a monomorphism. Hence $f : M \to N$ is a monomorphism, which finished the proof of the lemma.

To be able to conclude that a monomorphism $f : M \to N$ in $\text{mod}_I(\Lambda)$ gives a monomorphism $\overline{f} : \overline{M} \to \overline{N}$, we need some hypothesis on the algebra Λ.

PROPOSITION 1.2. Suppose that $\overline{\text{mod}(\Lambda)}$ has kernels, and let $f : M \to N$ be a map in $\overline{\text{mod}(\Lambda)}$. Then the following are equivalent.

(a) f is a monomorphism in $\text{mod}(\Lambda)$

(b) f is a monomorphism in $\text{mod}_I(\Lambda)$

(c) $\overline{f}$ is a monomorphism in $\overline{\text{mod}(\Lambda)}$.

PROOF: $\text{mod}_I(\Lambda)$ has kernels, coinciding with the kernels taken in $\text{mod}(\Lambda)$, since $\text{mod}_I(\Lambda)$ is closed under subobjects. For if $L \subset M$ has some injective summand, M would also have an injective summand. Hence (a) and (b) are equivalent. That (c) implies (b) follows from Lemma 1.1. To show that (b) implies (c), let $f : M \to N$ be a monomorphism in $\text{mod}_I(\Lambda)$. By assumption, $\overline{\text{mod}(\Lambda)}$ has kernels, so consider the exact sequence $0 \to \overline{L} \xrightarrow{\overline{g}} \overline{M} \xrightarrow{\overline{f}} \overline{N}$, with L in $\text{mod}_I(\Lambda)$. Consider the sequence $L \xrightarrow{g} M \xrightarrow{f} N$. Since $\overline{g}$ is mono, $g : L \to M$ is mono by Lemma 1.1. Hence fg is mono. Since $\overline{fg} = 0$, we conclude by Lemma 1.1 that $fg = 0$. Since f is not zero, g must be zero. Hence $L = 0$, since $L \xrightarrow{0} M$ is mono. We then conclude that $\overline{L} = 0$, hence that $0 \to \overline{M} \to \overline{N}$ is exact.

As an immediate consequence of Proposition 1.2, we get

PROPOSITION 1.3. Suppose $\overline{\text{mod}(\Lambda)}$ has kernels, and let M be an object in $\text{mod}_I(\Lambda)$. Then $\ell(M) = \ell(\overline{M})$.

We now apply this to the case of stably equivalent algebras Λ and Λ' such that $\overline{\text{mod}(\Lambda)}$ has kernels.

PROPOSITION 1.4. <u>Assume that</u> Λ <u>and</u> Λ' <u>are stably equivalent and that</u> $\overline{\mathrm{mod}(\Lambda)}$ <u>has kernels. The equivalence</u> $H : \overline{\mathrm{mod}(\Lambda)} \to \overline{\mathrm{mod}(\Lambda')}$ <u>has the property that if</u> M <u>is in</u> $\mathrm{mod}_I(\Lambda)$, <u>then</u> $\ell(M) = \ell(H(M))$.

PROOF: Immediate by Proposition 1.3.

We point out that since for Λ hereditary, $\overline{\mathrm{mod}(\Lambda)}$ is equivalent to $\mathrm{mod}_I(\Lambda)$, which has kernels, our results on the lenghts apply if one of the algebras is hereditary.

We are now ready to formulate some necessary conditions for Λ to be stably equivalent to an hereditary algebra.

THEOREM 1.5. <u>If an Artin algebra</u> Λ <u>is stably equivalent to some hereditary algebra</u> Γ, <u>the following is true.</u>

(a) <u>All non-simple indecomposable torsionless</u> Λ-<u>modules</u> M <u>are projective.</u>

(b) <u>If</u> M <u>is a</u> Λ-<u>module, and</u> $f : M \to S$ <u>an epimorphism, where</u> S <u>is a simple torsionless</u> Λ-<u>module, then</u> f <u>either splits or factors through an injective module.</u>

(c) <u>If</u> S <u>is a torsionless non-projective simple</u> Λ-<u>module, then</u> $S \subset E/\mathfrak{r}E$ <u>for some injective module</u> E.

PROOF: Let M be a non-simple non-injective indecomposable torsionless Λ-module. As before let $H : \mathrm{mod}_I(\Lambda) \to \mathrm{mod}_I(\Gamma)$ denote the map induced by the given equivalence of categories $H : \overline{\mathrm{mod}(\Lambda)} \to \overline{\mathrm{mod}(\Gamma)}$. By Proposition 1.5, $H(M)$ is not simple, and by Corollary 2.3 of Chapter II, $H(M)$ is also torsionless. Since Γ is hereditary, $H(M)$ is a non-simple indecomposable projective Γ-module. By Corollary 3.2 of Chapter II, M must also be an indecomposable non-simple projective module. To finish the proof of (a), let M be an indecomposable torsionless injective Λ-module. Since M is a submodule of a projective module, M is also a summand of a projective module, hence itself projective. This proves (a).

We go on to prove (b). Consider an epimorphism $f : M \to S$, where S is a simple torsionless (non-injective) Λ-module. By Proposition 1.4, $H(S)$ is simple, and by Corollary 2.3 of Chapter II, $H(S)$ is torsionless. Since Γ is hereditary, $H(S)$ is a simple projective Γ-module. Hence $H(\overline{f}) : H(M) \to H(S)$ is either split or zero, so the same is true for $\overline{f} : \overline{M} \to \overline{S}$. If $\overline{f} = 0$, then $f : M \to S$ factors through an injective module. If $f : \overline{M} \to \overline{S}$ splits, then $f : M \to S$ splits, since there is a representation equivalence $F : \mathrm{Mod}_I(\Lambda) \to \overline{\mathrm{mod}(\Lambda)}$ (analogous to the representation equivalence between $\mathrm{mod}_P(\Lambda)$ and $\underline{\mathrm{mod}(\Lambda)}$ mentioned in the introduction). This proves (b).

We use (b) to prove (c). Let S be a torsionless non-projective simple Λ-module, and let P be the projective cover of S. Consider $f : P \to S$, an epimorphism. Since f does not split, f factors through an injective module, hence we have a commutative diagram

$$\begin{array}{ccc} P & \xrightarrow{\ f\ } & S \\ & {\scriptstyle g}\searrow \quad \nearrow & \\ & E & \end{array}$$

Hence we have a map $h : E \to S$ from an injective module onto S, which means that $S \subset E/\mathfrak{R}E$.

We conclude this section by showing that if Λ and Λ' are stably equivalent Artin algebras which are stably equivalent to an hereditary algebra, then $L(\Lambda) = L(\Lambda')$, where L denotes the Loewy-length. The proof of this assertion depends on the following lemma concerning an additive category $\mathcal{A}$ satisfying

(i) Each object is uniquely the finite sum of indecomposable objects.

(ii) Each object A in $\mathcal{A}$ has finite length which we denote by $\ell(A)$.

(iii) Each indecomposable object B in $\mathcal{A}$ has a unique maximal subobject $A \xrightarrow{f} B$. That is, $\ell(B) = \ell(A) + 1$, and if $A' \xrightarrow{f'} B$ is another subobject of B satisfying $\ell(B) = \ell(A') + 1$, then there is an isomorphism $g: A \to A'$ such that $f = f'g$.

LEMMA 1.6. _Let_ $\mathcal{A}$ _be an additive category with the above properties. Then there is a unique function_ L _from the non-zero objects of_ $\mathcal{A}$ _to the positive integers satisfying_

(a) _If_ $B = \Sigma B_i$ (_finite direct sum_), _then_ $L(B) = \Sigma L(B_i)$.

(b) _If_ B _is simple, then_ $L(B) = 1$.

(c) _If_ B _has a unique maximal subobject_ $A \xrightarrow{f} B$, _then_ $L(B) = L(A) + 1$.

PROOF: We prove the uniqueness of L by induction on the length. Assume that L and L' both satisfy (a), (b), and (c). If B is simple, then $L(B) = 1 = L'(B)$. Assume that $L(B) = L'(B)$ if $\ell(B) < n$, and let $\ell(B) = n$. If B decomposes, $L(B) = L'(B)$ by (a) and the induction hypothesis. If B is indecomposable, then by (iii) there is a unique maximal subobject $A \xrightarrow{f} B$, and by (c) $L(B) = L(A) + 1 = L'(A) + 1 = L'(B)$.

We also prove the existence of L by induction on the length. If B is simple, we define $L(B) = 1$. Assume that $L(B)$ is defined for $\ell(B) < n$, and let $\ell(B) = n$. If B decomposes as $B = \Sigma B_i$, we define $L(B) = \Sigma L(B_i)$, which defines $L(B)$ uniquely by (i). If B is indecomposable, there is a unique maximal subobject $A \xrightarrow{f} B$ of B. Define $L(B) = L(A) + 1$. It is then clear that L satisfies (a), (b), and (c).

We now want to apply our lemma to the case where $\mathcal{A}$ is add(T), where T is the sum of (one copy of) each indecomposable torsionless non-injective Λ-module, and also to the case where $\mathcal{A}$ is add($\overline{T}$). Λ is supposed to be an Artin algebra stably equivalent to an hereditary algebra. (i) and (ii) are obviously satisfied for add(T) and add($\overline{T}$). (iii) holds for add(T) since each indecomposable torsionless Λ-module is simple or projective. We claim that (iii) also holds for add($\overline{T}$): Consider $\overline{P}$, where P is an indecomposable non-simple projective module. We want to show that $\overline{\mathfrak{r}P} \xrightarrow{\overline{i}} \overline{P}$ is a unique maximal subobject. It is clearly maximal. If $\overline{A} \xrightarrow{\overline{f}} \overline{P}$ was another maximal subobject, $A \xrightarrow{f} P$ would be a maximal subobject, so we would have the commutative diagram

$$\begin{array}{ccc} \mathfrak{r}P & \xrightarrow{i} & P \\ \wr\| & \nearrow_{f} & \\ \overline{A} & & \end{array}$$

hence the commutative diagram

$$\begin{array}{ccc} \overline{\mathfrak{r}P} & \xrightarrow{\overline{i}} & \overline{P} \\ \wr\| & \nearrow_{\overline{f}} & \\ \overline{A} & & \end{array}$$

which shows that $\overline{\mathfrak{r}P} \xrightarrow{\overline{i}} \overline{P}$ is unique.

We are now able to prove

PROPOSITION 1.7. <u>Assume that Λ and Λ' are stably equivalent Artin algebras and stably equivalent to an hereditary algebra. Let $H : \overline{\mathrm{mod}(\Lambda)} \to \overline{\mathrm{mod}(\Lambda')}$ be the given equivalence. If P is an indecomposable non-injective projective Λ-module, then</u> $L(P) = L(H(P))$.

PROOF: Let T be the sum of one copy of each of the indecomposable torsionless non-injective Λ-modules, and denote by L the unique functions as defined above on the two categories $\mathrm{add}(T)$ and $\mathrm{add}(\overline{T})$. Since L clearly coincides with the usual Loewy length on $\mathrm{add}(T)$, it is then sufficient to show that for a torsionless non-injective Λ-module P, $L(P) = L(\overline{P})$. We use induction on $\ell(P)$. If $\ell(P) = 1$, then $L(P) = L(\overline{P}) = 1$. Assume that we have equality for $\ell(P) < n$. Let $\ell(P) = n$. If P decomposes, we clearly get $L(P) = L(\overline{P})$. If P is indecomposable, then $L(P) = 1 + L(\mathfrak{r}P) = 1 + L(\overline{\mathfrak{r}P}) = L(\overline{P})$. This finishes the proof.

We finally are able to prove

THEOREM 1.8. *If* Λ *and* Λ' *are stably equivalent Artin algebras, stably equivalent to an hereditary algebra, then* $L(\Lambda) = L(\Lambda')$.

PROOF: We have already seen that our given equivalence $H : \overline{\text{mod}(\Lambda)} \to \overline{\text{mod}(\Lambda')}$ gives a one-one correspondence between the non-simple non-injective projective indecomposable modules of Λ and Λ'. By Proposition 1.7 $L(P) = L(H(P))$ if P is a (non-simple) non-injective projective indecomposable Λ-module.

Since by Corollary 1.4 of Chapter II, $E(S)$ is projective if and only if $E(H(S))$ is projective, and since $E(H(S))$ is indecomposable, we get a one-one correspondence between the indecomposable projective injective modules over Λ and Γ. Let $P = E(S)$ be projective, and consider $0 = \mathfrak{r}^n P \subset \mathfrak{r}^{n-1} P \subset \ldots \subset \mathfrak{r} P \subset P$. Since all $\mathfrak{r}^i P$ are indecomposable because they have simple socle, and all except possibly $S = \mathfrak{r}^{n-1} P$ are projective by (a), this is a composition series for P. Consider

$$0 \subset H(S) \subset H(\mathfrak{r}^{n-2} P) \subset \ldots \subset H(\mathfrak{r} P) \subset E(H(S)).$$

We conclude that $\ell(E(S)) = L(E(S)) \leq \ell(E(H(S))) = L(E(H(S)))$. Similarly, we get $L(E(H(S))) \leq L(E(S))$, hence equality. We have now set up a one-one correspondence between the indecomposable (non-simple) projectives for the two rings and shown that this correspondence preserves Loewy-length. Since $L(\Lambda) =$ the maximum of $L(P)$ for all indecomposable projective Λ-modules P, we conclude that $L(\Lambda) = L(\Lambda')$.

We remark that we could not hope to prove that the rings have the same lengths, since this property is not even preserved by Morita equivalence.

§2. SUFFICIENT CONDITIONS. In this section we will give sufficient conditions for an Artin algebra to be stably equivalent to an hereditary algebra. We have already shown that these conditions are necessary.

THEOREM 2.1. <u>Let Λ be an Artin algebra (with no semi-simple summand) with the following properties</u>

(1) <u>All indecomposable submodules of indecomposable projective modules are projective or simple.</u>

(2) <u>If S is a non-projective torsionless simple module, then $S \subset E/\mathfrak{R}E$ for some injective module E.</u>

<u>Then Λ is stably equivalent to an hereditary algebra.</u>

PROOF: Assume now that Λ is an algebra with the properties (1) and (2), although for some of the lemmas only one of the two is needed.

LEMMA 2.2. <u>Each indecomposable torsionless Λ-module is contained in an indecomposable projective Λ-module, hence is projective or simple.</u>

PROOF: Let M be an indecomposable torsionless module. Choose a projective module $P \supset M$, and let $P = P_1 \oplus \dots \oplus P_r$ be a decomposition of P into indecomposable projective modules. Consider the projection maps $p_i : P \to P_i$. If $p_i(M) \subset \operatorname{soc} P_i$ for $i = 1, \dots, r$, then $M \subset \operatorname{soc} P$. Since M is indecomposable, M is then simple.

Otherwise $p_i(M)$ is not semi-simple for some i. Let then N be an indecomposable summand of $p_i(M)$, which is not simple. N is then an indecomposable non-simple submodule of the indecomposable projective module P_i. By our assumption, N is then projective. Since M was indecomposable, the induced map from M onto N must be an isomorphism. Hence the lemma is proved.

LEMMA 2.3. <u>If $f : P \to S$ is an epimorphism onto a simple torsionless non-projective module S, with P projective, then f factors through an injective module.</u>

PROOF: Let $f : P \to S$ be an epimorphism, where P is projective and S is a torsionless non-projective simple Λ-module. Let E be such that $S \subset E/\mathfrak{R}E$.

Then there is an epimorphism $g : E \to S$, and since P is projective, there is a map $h : P \to E$ making the following diagram commute:

$$\begin{array}{ccc} P & \xrightarrow{f} & S \\ & {\scriptstyle h}\searrow \quad \nearrow{\scriptstyle g} & \\ & E & \end{array}$$

Hence f factors through an injective module, and the lemma is proved.

The following lemma is a generalization of Lemma 1.2 in Chapter III. There the ring was supposed to be hereditary. We point out that only condition (1) is needed here.

LEMMA 2.4. *If* M *is an indecomposable* Λ-*module and* $S \subset M$ *is a simple module such that* $E(S)$ *is projective, then* $M \subset E(S)$, *which implies that* $E(M)$ *is projective.*

PROOF: Consider the diagram

$$\begin{array}{ccccc} 0 & \to & S & \xrightarrow{i} & M \\ & & {\scriptstyle j}\downarrow & & \\ & & E(S) & & \end{array}$$

and let $f : M \to E(S)$ be a map which makes the diagram commute. Since $\operatorname{Im} f$ has simple socle, $\operatorname{Im} f$ is indecomposable. If $\operatorname{Im} f$ is simple, then $\operatorname{Im} f = S$. Then $i : S \to M$ would split, hence $S \cong M$, and we are done. Otherwise $\operatorname{Im} f$ is projective, which implies that $M \cong \operatorname{Im} f$, so $f : M \to E(S)$ is a monomorphism.

Before we go on we will introduce some more notation. We first point out that by property (1) there is only a finite number of indecomposable torsionless modules over Λ, since there is only a finite number of simple modules and a finite number of indecomposable projective modules. Let T be the sum of one copy of each of the non-injective indecomposable torsionless Λ-modules.

T_1 = the sum of the non-injective indecomposable torsionless Λ-modules M with $E(M)$ projective.

T_2 = the sum of the ones L with $E(L)$ not projective. $\mathcal{T} = \mathrm{add}(\overline{T})$, $\mathcal{T}_1 = \mathrm{add}(\overline{T}_1)$, $\mathcal{T}_2 = \mathrm{add}(\overline{T}_2)$, all full additive subcategories of $\overline{\mathrm{mod}(\Lambda)}$.

We have seen in Chapter I that $C \to \mathrm{Ext}^1(\ ,C)$ induces an equivalence of categories between $\overline{\mathrm{mod}(\Lambda)}$ and $\mathcal{J}$ = the full subcategory of injective objects $\hat{\mathcal{C}}_0(\Lambda)$. Since we have also proved in Chapter II, Proposition 2.2, that for C in $\mathrm{mod}_I(\Lambda)$, pd $\mathrm{Ext}^1(\ ,C) \leq 1$ if and only if C is a torsionless Λ-module, we get induced an equivalence between $\mathcal{T} = \mathrm{add}(\overline{T})$ and $\mathcal{T}'$ = the full subcategory of injective objects of projective dimension ≤ 1. Let $\mathcal{T}_1'$ be the full subcategory of projective injective objects of $\hat{\mathcal{C}}_0(\Lambda)$, $\mathcal{T}_2'$ the full additive subcategory generated by indecomposable injective objects of projective dimension one. Since by Proposition 1.2 of Chapter II, $\mathrm{Ext}^1(\ ,C)$ is projective if and only if $E(C)$ is projective, for C in $\mathrm{mod}_I(\Lambda)$, we also have equivalence of the categories $\mathcal{T}_1 = \mathrm{add}\ \overline{T}_1$ and $\mathcal{T}_1'$, and $\mathcal{T}_2 = \mathrm{add}\ \overline{T}_2$ and $\mathcal{T}_2'$. The aim of the next lemma is to show that $\mathcal{T}'$ decomposes as $\mathcal{T}' = \mathcal{T}_1' \times \mathcal{T}_2'$, or equivalently $\mathcal{T} = \mathcal{T}_1 \times \mathcal{T}_2$.

LEMMA 2.5. <u>Let</u> I_1 <u>and</u> I_2 <u>be indecomposable injective objects in</u> $\hat{\mathcal{C}}_0(\Lambda)$ <u>such that</u> pd $I_1 = 0$, pd $I_2 = 1$. <u>Then</u> $\mathrm{Hom}(I_1,I_2) = 0 = \mathrm{Hom}(I_2,I_1)$.

PROOF: Since pd $I_1 = 0$, $I_1 = \mathrm{Ext}^1(\ ,M)$, where M is an indecomposable module with $E(M)$ projective. Since pd $I_2 = 1$, $I_2 = \mathrm{Ext}^1(\ ,L)$, where L is an indecomposable module with $E(L)$ not projective. Hence we want to show that $\mathrm{Hom}(\overline{L},\overline{M}) = 0 = \mathrm{Hom}(\overline{M},\overline{L}) = 0$, for M and L indecomposable Λ-modules of the above types.

Assume that $f: L \to M$ is such that $\overline{f}: \overline{L} \to \overline{M}$ is not zero. Since L is indecomposable, and every non-simple indecomposable submodule of M is projective, either $\mathrm{Im}\, f$ is semi-simple, or f is a monomorphism. If $\mathrm{Im}\, f$ is semi-simple, f would factor through an injective module, so this would give $\overline{f} = 0$. If

$f : L \to M$ was a monomorphism, we would get a monomorphism $E(L) \to E(M)$. Hence $E(L)$ would be a summand of $E(M)$, which is a contradiction since $E(M)$ is projective and $E(L)$ is not. Hence $\mathrm{Hom}(\overline{L},\overline{M}) = 0$.

Suppose that $f : M \to L$ is such that $\overline{f} : \overline{M} \to \overline{L}$ is not zero. As above, we conclude that f must be a monomorphism. Hence we have an injection $E(M) \to E(L)$. Let $S \subset M \subset L$ be a simple module. By assumption, $E(S) \subset E(M)$ is projective. Hence L is an indecomposable module with a simple submodule S such that $E(S)$ is projective. By Lemma 2.4 we then conclude that $L \subset E(S)$, so that $E(L)$ is projective, a contradiction. This shows that $\mathrm{Hom}(\overline{M},\overline{L}) = 0$, and the proof is complete.

We now conclude that $\mathcal{A}'$ = {the category of injective objects of projective dimension ≤ 1} decomposes as $\mathcal{A}' = \mathcal{A}_1' \times \mathcal{A}_2'$.

LEMMA 2.6. __Let__ $(\,,C)$ __be an indecomposable projective object in__ $\hat{\mathcal{C}}_0$, __and__ $0 \to (\,,\underline{C}) \to E_0 \to E_1$ __a minimal injective copresentation in__ $\hat{\mathcal{C}}_0$. __Then either__ $\mathrm{pd}\, E_0 = \mathrm{pd}\, E_1 = 0$, __or every indecomposable summand of__ E_0 __and__ E_1 __has projective dimension__ 1.

PROOF: Let $(\,,\underline{C})$ be an indecomposable projective object of $\hat{\mathcal{C}}_0$, where C is an indecomposable non-projective Λ-module. Let P be the projective cover of C, and consider the exact sequence $0 \to K \to P \to C \to 0$. Then we have seen in Chapter I that $(\,,\underline{C})$ has a minimal injective copresentation

$$0 \to (\,,\underline{C}) \to \mathrm{Ext}^1(\,,K) \to \mathrm{Ext}^1(\,,P),$$

and K and P are torsionless Λ-modules. We decompose $\mathrm{Ext}^1(\,,K)$ and $\mathrm{Ext}^1(\,,P)$, each into two parts, to get

$$0 \to (\,,\underline{C}) \to \mathrm{Ext}^1(\,,M_0) \oplus \mathrm{Ext}^1(\,,L_0) \xrightarrow{g} \mathrm{Ext}^1(\,,M_1) \oplus \mathrm{Ext}^1(\,,L_1),$$

where $\mathrm{Ext}^1(\,,M_i)$ is in $\mathcal{A}_1'$, $\mathrm{Ext}^1(\,,L_i)$ in $\mathcal{A}_2'$. By Lemma 2.5,

$\mathrm{Hom}(\mathrm{Ext}^1(\ ,M_0),\ \mathrm{Ext}^1(\ ,L_1)) = 0 = \mathrm{Hom}(\mathrm{Ext}^1(\ ,L_0),\ \mathrm{Ext}^1(\ ,M_1))$. Hence g splits up as $g = g_1 \oplus g_2$, with $g_1 : \mathrm{Ext}^1(\ ,M_0) \to \mathrm{Ext}^1(\ ,M_1)$, $g_2 : \mathrm{Ext}^1(\ ,L_0) \to \mathrm{Ext}^1(\ ,L_1)$. It follows that $(\ ,\underline{C}) \cong \mathrm{Ker}\ g_1 \oplus \mathrm{Ker}\ g_2$, which implies that either $\mathrm{Ker}\ g_1$ or $\mathrm{Ker}\ g_2$ is $(\ ,\underline{C})$, since $(\ ,\underline{C})$ is indecomposable. Hence either $\mathrm{Ext}^1(\ ,M_i) = 0$ or $\mathrm{Ext}^1(\ ,L_i) = 0$, since the copresentation was supposed to be minimal. This proves the lemma.

This lemma shows that $\mathcal{P} = \{$the full subcategory of projective objects of $\hat{\mathcal{C}}_0\}$ decomposes as $\mathcal{P} = \mathcal{P}_1 \times \mathcal{P}_2$, where $\mathcal{P}_1$ consists of the projective objects having minimal injective copresentations with injective projectives, and $\mathcal{P}_2$ consists of the projective objects where the two first terms in an injective resolution have all summands of projective dimension 1. And if there was a non-zero map $f : P_1 \to P_2$, where $P_1 \in \mathcal{P}_1$, $P_2 \in \mathcal{P}_2$, we would get induced a non-zero map from $E(P_1)$ to $E(P_2)$, a contradiction.

Now it can be seen that this decomposition $\mathcal{P} = \mathcal{P}_1 \times \mathcal{P}_2$ induces a decomposition $\hat{\mathcal{C}}_0 = \mathcal{C}_1 \times \mathcal{C}_2$, where the indecomposable objects in $\mathcal{C}_i$ are the indecomposable objects having minimal projective presentations using projectives from $\mathcal{P}_i$.

We want to find a candidate for the hereditary algebra which should be stably equivalent to Λ. We would like to find a construction which for Λ hereditary would give Λ itself back again. The natural thing in that case is to look at the indecomposable projective modules. If we consider the endomorphism ring of their sum in $\overline{\mathrm{mod}(\Lambda)}$, we get an hereditary ring, but in general not the one we started with. However, by our structure theory for hereditary algebras, we can get Λ back again by the operation α^{-1} mentioned at the end of the previous chapter. The analogous ring to look at for Λ not hereditary is $\Gamma = \mathrm{End}(\overline{T})^{op}$, where T is the sum of one copy or each of the torsionless non-injective indecomposable Λ-modules.

We want to show that Γ is hereditary, and finally that $\alpha^{-1}(\Gamma)$ is stably equivalent to Λ.

LEMMA 2.7. $\Gamma = \text{End}(\overline{T})^{op}$ <u>is hereditary</u>.

PROOF: We will consider the objects of $\text{mod}(\Gamma)$ as the coherent functors $\text{add}\hat{(}\overline{T})$ from $\text{add}(\overline{T})$ to Ab. We know that Γ is an Artin algebra, since Λ is, and T is a finitely generated Λ-module. Hence $\text{mod}(\Gamma)$ has projective covers, and it is sufficient to show that $\text{pd}\, F \leq 1$ for all simple objects F [1].

Let $(\ ,\overline{A}) \xrightarrow{f'} (\ ,\overline{B}) \to F \to 0$ be a minimal projective presentation of F. Let $f : A \to B$ be a map inducing $f' : (\ ,\overline{A}) \to (\ ,\overline{B})$.

If B is a simple Λ-module, then f factors through an injective Λ-module. For B is torsionless, and A is a sum of indecomposable projective Λ-modules. If A had a simple summand A_1, $A_1 \xrightarrow{g} B$ would be zero, in which case we get a contradiction to the minimality of the resolution, or $g : A_1 \to B$ an isomorphism, which would give $g' : (\ ,\overline{A}) \to (\ ,\overline{B})$ an epimorphism, also a contradiction. We can now conclude that $F \cong (\ ,\overline{B})$ is projective if B is simple.

If B is not simple, B is an indecomposable projective Λ-module. We then claim that the sequence $0 \to (\ ,\overline{\mathfrak{R}B}) \xrightarrow{\overline{i}'} (\ ,\overline{B}) \to F \to 0$ is exact, hence showing that $\text{pd}\, F = 1$ in this case.

We first show that $\overline{i} : \overline{\mathfrak{R}B} \to \overline{B}$ is a monomorphism. So assume there is a map $\overline{g} : \overline{X} \to \overline{B}$ such that in $\overline{X} \xrightarrow{\overline{g}} \overline{B} \xrightarrow{\overline{i}} B$, $\overline{i}\,\overline{g}$ is zero. Hence there is a commutative diagram of Λ-modules, with E injective

$$\begin{array}{ccccc} X & \xrightarrow{g} & \mathfrak{R}B & \xrightarrow{i} & B \\ & \searrow & & \nearrow_{h} & \\ & & E & & \end{array}$$

If $h : E \to B$ is onto, h splits, since B is projective, and B would be a summand of E, hence itself an injective module, a contradiction. Hence $h(E) \subset \mathfrak{R}B$, which means that $\overline{g} : \overline{X} \to \overline{\mathfrak{R}B}$ is zero. We conclude that $\overline{i}$ is mono. Hence $\overline{i}' : (\ ,\overline{\mathfrak{R}B}) \to (\ ,\overline{B})$ is mono.

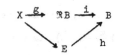

If we can show that $\mathrm{Coker}(\bar{i}')$ is simple, it must be F, which is the top of $(\ ,\bar{B})$. Consider the exact sequence $0 \to (\ ,\overline{\mathfrak{R}B}) \xrightarrow{\bar{i}'} (\ ,\bar{B}) \xrightarrow{j'} \mathrm{Coker}(\bar{i}') \to 0$. To show that $\mathrm{Coker}(\bar{i}')$ is simple it suffices to show that for any map $X \xrightarrow{h'} (\ ,\bar{B})$, where $\mathrm{Im}\, h' \not\subset (\ ,\overline{\mathfrak{R}B})$, $j'h' : X \to \mathrm{Coker}(\bar{i}')$ is epi. It is clearly enough to consider projective objects $X = (\ ,\bar{C})$. Let $h : C \to B$ induce $h' : (\ ,\bar{C}) \to (\ ,\bar{B})$. If $\mathrm{Im}\, h \subset \mathfrak{R}B$, then $\mathrm{Im}\, h' \subset (\ ,\overline{\mathfrak{R}B})$. Hence $\mathrm{Im}\, h \not\subset \mathfrak{R}B$, so $h : C \to B$ is onto. Hence h splits, and so $h' : (\ ,\bar{C}) \to (\ ,\bar{B})$ splits, and is consequently epi. Hence also $j'h' : (\ ,\bar{C}) \to \mathrm{Coker}(\bar{i}')$ is epi. This shows that $\mathrm{Coker}(\bar{i}')$ is simple, which finishes the proof that $0 \to (\ ,\overline{\mathfrak{R}B}) \to (\ ,\bar{B}) \to F \to 0$ is exact. We conclude that $\Gamma = \mathrm{End}(\bar{T})^{op}$ is hereditary.

LEMMA 2.8. <u>The objects of</u> $\mathrm{Comod}\ \mathcal{A}' \subset \hat{\mathcal{C}}_0$ <u>are either projective or injective in</u> $\hat{\mathcal{C}}_0$, <u>where</u> $\mathcal{A}'$ <u>is the full additive subcategory of</u> $\hat{\mathcal{C}}_0$ <u>generated by the injective objects of projective dimension at most one.</u>

PROOF: Let F be an indecomposable non-injective object in $\hat{\mathcal{C}}_0$, and let $0 \to F \to \mathrm{Ext}^1(\ ,A) \xrightarrow{g'} \mathrm{Ext}^1(\ ,B)$ be a minimal injective copresentation. Since F is not injective, g' is not zero, and is induced by $g : A \to B$ in $\mathrm{mod}_I(\Lambda)$, $\bar{g} : \bar{A} \to \bar{B}$ in $\overline{\mathrm{mod}(\Lambda)}$. g cannot be decomposed, since F is indecomposable, and the injective copresentation is minimal. Consider the following resolution of $G = \mathrm{Coker}(\bar{g}')$ in $\mathrm{mod}(\Gamma)$: $(\ ,\bar{A}) \xrightarrow{\bar{g}'} (\ ,\bar{B}) \to G \to 0$. Since $\mathrm{mod}(\Gamma)$ has projective covers, and $\bar{g}'$ is indecomposable, this must be a minimal projective presentation for G. Hence $0 \to (\ ,\bar{A}) \xrightarrow{\bar{g}'} (\ ,\bar{B})$ is exact since $\mathrm{gl.\,dim}(\Gamma) \leq 1$, so also $0 \to \bar{A} \xrightarrow{\bar{g}} \bar{B}$ is exact. By Lemma 1.1 $0 \to A \xrightarrow{g} B$ is also exact. Consider then the exact sequence $0 \to A \xrightarrow{g} B \to C \to 0$, where $C \to \mathrm{Coker}\, g$. If we can show that B is projective, we would have a minimal injective copresentation for $(\ ,\underline{C})$ as follows: $0 \to (\ ,\underline{C}) \to \mathrm{Ext}^1(\ ,A) \xrightarrow{g'} \mathrm{Ext}^1(\ ,B)$. This would imply $F \cong (\ ,\underline{C})$, hence that F is projective.

Since B is torsionless, we need only prove that B has no simple summands. So assume to the contrary that $B \cong B' \oplus S$, where S is a simple Λ-module. Consider $g' : \mathrm{Ext}^1(\ ,A) \to \mathrm{Ext}^1(\ ,B) \oplus \mathrm{Ext}^1(\ ,S)$. The only way to have a non-zero map $g_1' : \mathrm{Ext}^1(\ ,A) \to \mathrm{Ext}^1(\ ,S)$ would be to have isomorphism on a summand of $\mathrm{Ext}^1(\ ,A)$ with $\mathrm{Ext}^1(\ ,S)$. But this would contradict the minimality of the resolution, and so would also $g_1' = 0$. Hence B must be projective, and we are done.

LEMMA 2.9. Comod $\mathcal{A}'$ *is equivalent to* $\mathrm{mod}(\Gamma)$, *where* $\Gamma = \mathrm{End}(\overline{T})^{op}$. *And the injective objects of* Comod $\mathcal{A}'$ *coincide with the objects of* Comod $\mathcal{A}'$ *which are injective in* $\hat{\mathcal{C}}_0$.

PROOF: We know that the natural functor $V : \mathcal{A}' \to \hat{\mathcal{A}}' \sim \mathrm{mod}(\Gamma)$ induces an equivalence between $\mathcal{A}'$ and $\mathcal{P}(\Gamma)$ = {the full subcategory of projective objects in $\mathrm{mod}(\Gamma)$}. Hence we get an equivalence of categories $\alpha : \mathrm{Comod}\ \mathcal{A}' \to \mathrm{Comod}\ \mathcal{P}(\Gamma)$. Further, there is a duality $\beta' : \mathcal{P}(\Gamma) \to \mathcal{P}(\Gamma^{op})$ given by $\beta'(P) = \mathrm{Hom}_\Gamma(P,\Gamma) = P^*$, which induces a duality $\beta : \mathrm{Comod}\ \mathcal{P}(\Gamma) \to \mathrm{Mod}(\mathcal{P}(\Gamma^{op}) = \mathrm{mod}(\Gamma^{op})$ (see Chapter I). And finally, $D : \mathrm{mod}(\Gamma^{op}) \to \mathrm{mod}(\Gamma)$, given by $D(M) = \mathrm{Hom}_R(M, E_R(R/\mathfrak{r})$ where R is the ground ring, is a duality. Hence $D\beta\alpha : \mathrm{Comod}\ \mathcal{A}' \to \mathrm{mod}(\Gamma)$ is an equivalence of categories.

The statement about the injective objects follows from the fact that the natural functor $U : \mathrm{Comod}\ \mathcal{A}' \subset \hat{\mathcal{C}}_0$ is a left exact embedding. Hence if A,B are in Comod $\mathcal{A}'$, $0 \to A \to B$ is exact in Comod $\mathcal{A}'$ if and only if $0 \to A \to B$ is exact in $\hat{\mathcal{C}}_0$. It is then easy to see that X in Comod $\mathcal{A}'$ is injective if and only if X is injective in $\hat{\mathcal{C}}_0$.

The next lemma has a similar proof as the analogous lemma for hereditary algebras.

LEMMA 2.10. $\Gamma_1 = \mathrm{End}(\overline{T}_1)^{op}$ *is a finite product of lower triangular matrix* rings over a division ring, $Tr_{m_i}(D_i)$, where T_1 is the sum of one copy of each of the indecomposable torsionless Λ-modules M with $E(M)$ projective.

PROOF: If M is an indecomposable Λ-module with $E(M)$ projective, then M has simple socle. For let $S \subset M$ be a simple submodule. Since $E(S)$ is projective, we conclude by Lemma 2.4 that $M \subset E(S)$. Now $E(S)$ is uniserial, since all $\mathfrak{r}^i E(S)$ are indecomposable simple or projective. Hence we have a chain $S = P_1 \subset P_2 \subset \dots \subset P_n \subset E(S)$ of all the indecomposable Λ-modules with socle S. Let Q be an indecomposable Λ-module with $\operatorname{soc} Q = T \not\approx S$. If $\operatorname{Hom}(\overline{P_i},\overline{Q}) \neq 0$ we must have an inclusion $P_i \subset Q$, which is impossible since the socles are different. For if $f : P_i \to Q$ is a map such that $\operatorname{Im} f = T \subset Q$, then f would factor through an injective module, since P_i is projective (Lemma 2.3) and T is simple. Hence $\operatorname{Hom}(\overline{P_i},\overline{Q}) = 0$, and similarly $\operatorname{Hom}(\overline{Q},\overline{P_i}) = 0$.

We then consider $\Omega = \operatorname{End}(\overline{P_1} \oplus \dots \oplus \overline{P_n})^{op}$. We observe that $\operatorname{Hom}(\overline{P_i},\overline{P_j}) \neq 0$ only if $P_i \subset P_j$. If $P_i \subset P_j$, then $\operatorname{Hom}(\overline{P_i},\overline{P_j}) \simeq \operatorname{Hom}(P_i,P_j)$, since none of the inclusion maps factor through an injective module. Hence $\Omega \simeq \operatorname{End}_\Lambda(P_1 + \dots + P_n)^{op}$. The proof now continues like the proof of Proposition 1.3 in Chapter III, using that $E \supset P_n$ is an injective module.

We are now ready to finish the proof of the theorem. We have $\operatorname{Comod} \mathcal{a}'$ equivalent to $\operatorname{mod}(\Gamma)$, a decomposition $\operatorname{Comod} \mathcal{a}' = \operatorname{Comod} \mathcal{a}_1' \times \operatorname{Comod} \mathcal{a}_2'$, and a corresponding decomposition $\operatorname{mod}(\Gamma) = \operatorname{mod}(\Gamma_1) \times \operatorname{mod}(\Gamma_2)$. $\operatorname{Comod} \mathcal{a}_2' \subset \mathcal{C}_2$ consists of the projective objects of $\mathcal{C}_2$ and the indecomposable injective objects of projective dimension one. Hence $\operatorname{mod}_I(\Gamma_2)$ is equivalent to the category of projective objects of $\mathcal{C}_2$.

Further, $\operatorname{Comod} \mathcal{a}_1'$ consists of the projective objects of $\mathcal{C}_1$, hence $\operatorname{mod}(\Gamma_1)$ is equivalent to the category of projective objects of $\mathcal{C}_1$. By Lemma 2.10, Γ_1 is a finite product of rings $Tr_{m_i}(D_i)$. Hence we know that $\operatorname{mod}(\alpha^{-1}(\Gamma_1))$ is equivalent to $\operatorname{mod}(\Gamma_1)$. Let $\Gamma' = \alpha^{-1}(\Gamma_1) \times \Gamma_2 = \alpha^{-1}(\Gamma) =$

$\alpha^{-1}(\text{End}(\overline{T})^{\text{op}})$. Then we have established that $\overline{\text{mod}(\Gamma')}$ is equivalent to the category of projective objects in $\hat{C}_0$, and this category is again equivalent to $\underline{\text{mod}(\Lambda)}$, which is equivalent to $\overline{\text{mod}(\Lambda)}$. Hence $\overline{\text{mod}(\Gamma')}$ and $\overline{\text{mod}(\Lambda)}$ are equivalent. This finishes the proof of the theorem.

We will discuss the case $\mathfrak{R}^2 = 0$, in the last chapter, giving a different approach, and an easier way of constructing the hereditary ring in that special case. We here show as an easy consequence of Theorem 2.1 that all algebras with $\mathfrak{R}^2 = 0$ are stably equivalent to some hereditary algebra.

COROLLARY 2.11 . **An Artin algebra** Λ **with** $\mathfrak{R}^2 = 0$ **is stably equivalent to an hereditary algebra**.

PROOF: We verify the conditions (1) and (2) if $\mathfrak{R}^2 = 0$. If M is an indecomposable submodule of an indecomposable projective module P , and $M \not\cong P$, then $M \subset \mathfrak{R}P$, hence M is simple.

Let S be a simple torsionless non-projective module, and let P be the projective cover of S . Consider $E(P)$. $\mathfrak{R}P = \text{soc } P = \text{soc } E(P) = \mathfrak{R} E(P)$. Hence $S = P/\mathfrak{R}P \subset E(P)/\mathfrak{R} E(P)$.

§3 . EXAMPLES . In this section we will give some examples of algebras which are stably equivalent to hereditary algebras, where $\mathfrak{R}^2 \neq 0$. We will show how to construct new examples from given ones, and we will also show that there are examples of algebras satisfying (1) but not (2), and also (2) but not (1).

EXAMPLE 3.1 . Let k be a field, and consider

$$\Lambda = \left\{ \begin{pmatrix} \alpha & & 0 \\ a_1 a_2 & \\ a_3 a_4 & \alpha \end{pmatrix} \quad , \ a_i, \ \alpha \in k \right\}$$

Λ has the idempotents $e_1 = \begin{pmatrix} 1&0&0 \\ 0&0&0 \\ 0&0&1 \end{pmatrix}$ and $e_2 = \begin{pmatrix} 0&0&0 \\ 0&1&0 \\ 0&0&0 \end{pmatrix}$. Let $P_1 = \Lambda e_1$, $P_2 = \Lambda e_2$, $S_1 = P_1/\mathfrak{R} P_1$, $S_2 = P_2/\mathfrak{R} P_2$. The exact sequences $0 \to S_1 \to P_2 \to S_2 \to 0$ and $0 \to P_2 \to P_1 \to S_1 \to 0$ show that gl. dim. $\Lambda = 2$. The submodules of the indecomposable projective modules P_1 and P_2 are P_1, P_2, S_1, hence they are all projective or simple. And the only torsionless non-projective simple module S_1 occurs in the topy of P_1, which is injective. Hence both (1) and (2) are satisfied. Now $\operatorname{End}(\overline{P_1 \oplus P_2 \oplus S_1})^{op} = \operatorname{End}(\overline{P_2} \oplus \overline{S_1})^{op} = \begin{pmatrix} k & 0 \\ k & k \end{pmatrix}$. Hence Λ is stably equivalent to $\Gamma = \begin{pmatrix} k & & 0 \\ k & k & \\ k & k & k \end{pmatrix}$.

Similarly $\Lambda = \left\{ \underbrace{\begin{pmatrix} \alpha & & & \\ a_1 & a_2 & & 0 \\ \vdots & & \ddots & \\ \vdots & \cdots & & \alpha \end{pmatrix}}_{n}, \; a_i, \alpha \in k \right\}$ has gl. dim $= 2$ and is stably equivalent to $\Gamma = Tr_n(k)$.

Let Λ be an hereditary algebra, S a simple non-projective Λ-module. The following result shows how to construct examples of algebras with properties (1) and (2), and at the same time gives examples to show that (2) cannot be left out.

PROPOSITION 3.2. *Let Λ be an hereditary algebra, S a simple non-projective Λ-module. Then $\Gamma = \operatorname{End}(\Lambda \oplus S)^{op}$ is an algebra with gl. dim $\Gamma = 2$ with property (1). And Γ has property (2) if and only if S is an injective Λ-module.*

PROOF: We regard the objects of $\operatorname{mod}(\Gamma)$ as coherent functors from $\operatorname{add}(\Lambda \oplus S)$ to Ab. Let F be a simple object. If F is the top of $(\ ,P)$, with P indecomposable projective, $0 \to (\ ,\mathfrak{R}P) \xrightarrow{i'} (\ ,P) \xrightarrow{j'} F \to 0$ is an exact sequence. To show this, we want to show that Coker i' is simple. Let $h' : (\ ,Q) \to (\ ,P)$ be a map with $\operatorname{Im} h' \not\subset (\ ,\mathfrak{R}P)$. We then need to show that $j'h' : (\ ,Q) \to \operatorname{Coker} i'$

is epi. Let $h: Q \to P$ induce $h': (\ ,Q) \to (\ ,P)$. Then $\operatorname{Im} h \not\subset \mathfrak{r}P$, hence $h: Q \to P$ splits, so h' splits, and we conclude that $j'h'$ is epi.

If F is the top of $(\ ,S)$, we claim that $0 \to (\ ,P) \xrightarrow{i'} (\ ,P) \xrightarrow{\ell'} (\ ,S) \xrightarrow{j'} F \to 0$ is exact. Since $0 \to \mathfrak{r}P \to P \to S \to 0$ is exact, hence also $0 \to (\ ,\mathfrak{r}P) \to (\ ,P) \to (\ ,S)$, we need only show that $\operatorname{Coker} \ell'$ is simple. So consider again $h': (\ ,Q) \to (\ ,S)$, where Q is an indecomposable projective Λ-module or $Q = S$, such that h' does not factor through ℓ'. If $h: Q \to S$ is not zero, $Q = P$, or $Q = S$. In the first case h' factors through ℓ', and in the second case $(\ ,S) \xrightarrow{=} (\ ,S) \to \operatorname{Coker} j'$ is epi. This shows that $\operatorname{Coker} j'$ is simple. This discussion gives gl. dim $\Gamma = 2$.

The subobjects of $(\ ,P)$ for an indecomposable projective Λ-module P are the $(\ ,Q)$, where $Q \subset P$. We have seen that $\ell(\ ,S) = 2$, hence a proper submodule of $(\ ,S)$ is simple.

We assume that S is injective, and want to show that property (2) is satisfied for Γ. Let G be a simple which is the top of $(\ ,P)$ where P is projective and assume $G \subset (\ ,Q)$. If Q is projective, there is a map $f: P \to Q$ which must be a monomorphism. Hence $(\ ,P) \to G \to 0$ is mono, so G is projective. Hence the only torsionless non-projective module is the simple module G contained in $(\ ,S)$. We then have

$$\begin{array}{ccccccccccc} 0 & \to & (\ ,\mathfrak{r}P) & \to & (\ ,P) & & \to & & (\ ,S) & \to F \to 0 \\ & & & & & \searrow & & \nearrow & & \\ & & & & & & G & & & \\ & & & & & \nearrow & & \searrow & & \\ & & & & 0 & & & & 0 & \end{array}$$

$P \xrightarrow{f} S$ gives rise to a commutative diagram

$$\begin{array}{ccc} P & \longrightarrow & S \\ & \searrow \quad \nearrow & \\ & E(P) & \end{array} \quad ,$$

since S is injective. Consider

$$\begin{array}{ccc} (\ ,P) & \longrightarrow & (\ ,S) \\ & \searrow \qquad \nearrow g' & \\ & (\ ,E(P)) & \end{array} .$$

Since $E(P)$ is an injective Λ-module, $(\ ,E(P))$ is an injective Γ-module [4]. To show that G is in the top of $(\ ,E(P))$, we need only show that $g' : (\ ,E(P)) \to (\ ,S)$ is not epi. If it was, g' would split, so S would be a summand of $E(P)$, hence $S \subset P$, a contradiction.

Assume now that S is not injective. If $(\ ,S)$ was in the top of some $(\ ,E)$ where E is injective, the map $g' : (\ ,E) \to (\ ,S)$ would be induced by some non-zero map $g : E \to S$. But since Λ is hereditary, this would imply that S be injective. This finishes the proof of the proposition.

For the case of S being injective, we can describe the hereditary algebra stably equivalent to Γ. $\mathrm{End}(\overline{T})^{op} = \mathrm{End}(\overline{(\ ,\Lambda) \oplus G \oplus (\ ,S)})^{op} = \mathrm{End}(\overline{(\ ,\Lambda) \oplus G})^{op} = \alpha(\Lambda) \times D$, where $D = (\mathrm{End}\ G)^{op}$. Then Γ is stably equivalent to the hereditary algebra $\alpha^{-1}(\alpha(\Lambda) \times D) = \Lambda \times \alpha^{-1}(D) = \Lambda \times \begin{pmatrix} D & 0 \\ D & D \end{pmatrix}$.

Letting Λ be an hereditary algebra and S a simple injective module, Proposition 3.2 shows how to construct examples of algebras with properties (1) and (2). And if we choose an hereditary algebra Λ and a simple module S which is neither projective nor injective, then $\mathrm{End}(\Lambda + S)^{op}$ satisfies (1), but not (2). An example of such an algebra Λ is $\begin{pmatrix} k & & 0 \\ k & k & \\ k & k & k \end{pmatrix}$, where $S = \Lambda e_2 / \mathfrak{R} e_2$, $e_2 = \begin{pmatrix} 0 & 0 & 0 \\ 0 & 1 & 0 \\ 0 & 0 & 0 \end{pmatrix}$.

We provide two examples of a different kind, where one ring satisfies (1) but not (2), the other one (2) but not (1).

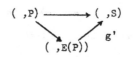

EXAMPLE 3.3. $\Lambda = \left\{ \begin{pmatrix} a_1 & 0 \\ a_2\alpha & \\ a_3 a_4 \alpha & \end{pmatrix},\ a_i,\ \alpha \in k,\ \text{a field} \right\}$ satisfies condition (1),

but not (2) : Let $e_1 = \begin{pmatrix} 1&0&0 \\ 0&0&0 \\ 0&0&0 \end{pmatrix}$, $e_2 = \begin{pmatrix} 0&0&0 \\ 0&1&0 \\ 0&0&1 \end{pmatrix}$. $P_1 = \Lambda e_1$, $P_2 = \Lambda e_2$, $S_1 = \Lambda e_1/\mathfrak{R} e_2$, $S_2 = \Lambda e_2/\mathfrak{R} e_2$. Then the submodules of P_1 and P_2 are P_1, P_2 and S_2 , hence all simple or projective. Further S_2 is a torsionless non-projective module which is not in the top of any injective module.

EXAMPLE 3.4 . $\Lambda = \left\{ \begin{pmatrix} \alpha & & 0 \\ a_1\alpha & & \\ a_2 & a_3 & a_4 \end{pmatrix} , a_i , \alpha \in k , \text{ a field} \right\}$ satisfies condition (2), but not (1) : Let $e_1 = \begin{pmatrix} 1&0&0 \\ 0&1&0 \\ 0&0&0 \end{pmatrix}$, $e_2 = \begin{pmatrix} 0&0&0 \\ 0&0&0 \\ 0&0&1 \end{pmatrix}$. Here $P_1 = \Lambda e_1$ has an indecomposable submodule which is neither projective nor simple. And (2) holds, since S_2 is projective and S_1 is not torsionless, and there are no other simple modules.

We next show that, starting with Λ satisfying (1) and (2), we can construct another ring Γ with the properties (1) and (2), such that gl. dim $\Gamma \leq 2$.

PROPOSITION 3.5 . Assume Λ has the properties (1) and (2) . Then $\Gamma = \text{End}_{\Gamma}(T)^{op}$, where T is the sum of one copy of each of the indecomposable torsionless modules, satisfies (1) and (2) , and gl. dim $\Gamma \leq 2$.

PROOF: We consider the Γ-modules as coherent functors on add T . If the simple module F is the top of $(,P)$, where P is a projective Λ-module, we get as before that $0 \to (,\mathfrak{R}P) \to (,P) \to F \to 0$ gives a projective resolution for F . If F is the top of $(,S)$ with S simple, we get that $0 \to (,\mathfrak{R}P) \to (,P) \to (,S) \to F \to 0$ is exact, where P is the projective cover of S , in a similar way as in the previous proposition. We conclude that gl. dim $\Gamma \leq 2$. To show that Γ satisfies property (1), let $(,P)$ be an indecomposable projective module, and $F \subset (,P)$ an indecomposable submodule. F is given as $\text{Im } f'$, where $f' : (,Q) \to (,P)$ is induced by $f : Q \to P$ ((,Q) is the projective cover of F). If $f : Q \to P$ is mono, $F \cong (,Q)$ is projective. Otherwise one can see that $\operatorname{Im} f$ must be semi-simple $= S_1 \oplus \dots \oplus S_r$. Hence $F \subset (\ ,S_1) \oplus \dots \oplus (\ ,S_r)$. Since F is indecomposable and each $(\ ,S_i)$ has length 2, we conclude that $F \subset (\ ,S_i)$ for some i, so $F = (\ ,S_i)$, or is simple.

As in the proof of the previous proposition, one can see that the torsionless non-projective simple modules are the simple submodules G of $(\ ,S)$, where S is a torsionless non-projective module. Consider then

$$\begin{array}{ccccccccc} 0 \to (\ ,P) & \to & (\ ,P) & & \to & & (\ ,S) \to F \to 0 \\ & & & \searrow & & \nearrow & & & \\ & & & & G & & & & \\ & & & \nearrow & & \searrow & & & \\ & & 0 & & & & 0 & & \end{array}$$

It is not too hard to see that $(\ ,P) \to G$ factors through an injective module, using that $P \to S$ does.

This construction will in general give algebras with larger Loewy-length. So starting with algebras of $\mathfrak{R}^2 = 0$, which we know satisfy the hypothesis, we can construct several examples with $\mathfrak{R}^2 \neq 0$.

We mention that we can get back to Λ by taking the (opposite) endomorphism ring of an appropriate projective Γ-module. Hence we can get any algebra with the properties (1) and (2) as (opposite) endomorphism rings of projective modules over rings of global dimension at most 2 with properties (1) and (2).

CHAPTER V

In this chapter we want to give a different approach to the case where $\mathfrak{R}^2 = 0$, now mostly considering Artin rings rather than Artin algebras. We show that each Artin ring Ω with $\mathfrak{R}^2 = 0$ is (projectively) stably equivalent to the hereditary ring $\Gamma = \begin{pmatrix} \Omega/\mathfrak{R} & 0 \\ \mathfrak{R} & \Omega/\mathfrak{R} \end{pmatrix}$, by defining a natural functor $F : \operatorname{mod}(\Omega) \to \operatorname{mod}(\Gamma)$. Hence this gives an easy way of constructing an hereditary ring in the same stable equivalence class. By describing how to get an hereditary ring with no semi-simple ring summands, and using the uniqueness of the hereditary ring for Artin algebras, we can formulate a necessary and sufficient condition for two Artin algebras Ω and Ω' with $\mathfrak{R}^2 = \mathfrak{R}'^2 = 0$ to be stably equivalent: Consider $\mathfrak{R}$ as an Ω_2-Ω_1-bimodule, where $\Omega_2 = \Omega/$ left ann $\mathfrak{R}$, $\Omega_1 = \Omega/$ right ann $\mathfrak{R}$. Then Ω and Ω' have equivalent stable categories if and only if ${}_{\Omega_2}\mathfrak{R}_{\Omega_1}$ and ${}_{\Omega'_2}\mathfrak{R}'_{\Omega'_1}$ are isomorphic bimodules, in the sense that there are ring isomorphisms $f : \Omega_2 \to \Omega'_2$, $g : \Omega_1 \to \Omega'_1$ and a group isomorphism $h : \mathfrak{R} \to \mathfrak{R}'$ such that $h(ar) = f(a)h(r)$, $h(rb) = h(r)g(b)$, where $r \in \mathfrak{R}$, $a \in \Omega_2$, $b \in \Omega_1$.

§1. TRIANGULAR MATRIX RINGS. In this section we give some background on modules over triangular matrix rings.

For an Artin ring Λ and a two-sided Λ-module M, we will consider the trivial extension $\Lambda \times M$, where multiplication is given by $(\lambda,m)\cdot(\lambda',m') = (\lambda\lambda', \lambda m' + m\lambda')$, and addition by $(\lambda,m) + (\lambda',m) = (\lambda + \lambda', m + m')$. As a special case of this we will consider, for two rings Λ_1 and Λ_2, and a Λ_2-Λ_1-bimodule M, the triangular matrix ring $\Gamma = \begin{pmatrix}\Lambda_1 & 0\\ M & \Lambda_2\end{pmatrix}$. Here the multiplication is given by $\begin{pmatrix}\lambda_1 & 0\\ m & \lambda_2\end{pmatrix}\cdot\begin{pmatrix}\lambda'_1 & 0\\ m' & \lambda_2\end{pmatrix} = \begin{pmatrix}\lambda_1\lambda'_1 & 0\\ m\lambda'_1 + \lambda_2 m' & \lambda_2\lambda'_2\end{pmatrix}$,

and addition by $\begin{pmatrix}\lambda_1 & 0\\ m & \lambda_2\end{pmatrix} + \begin{pmatrix}\lambda'_1 & 0\\ m' & \lambda'_2\end{pmatrix} = \begin{pmatrix}\lambda_1 + \lambda'_1 & 0\\ m + m' & \lambda_2 + \lambda'_2\end{pmatrix}$. If we write $\Lambda = \Lambda_1 \times \Lambda_2$, and consider M as a two-sided Λ-module in the natural way, then $\Gamma = \Lambda \times M$.

It will be convenient to consider the (left) Λ-modules as triples (A,B,f), where A is a (left) Λ_1-module, B a Λ_2-module and $f : M \otimes A \to B$ a Λ_2-homomorphism. The maps between two Γ-modules (A,B,f) and (A',B',f') are then pairs (g_1,g_2), where g_1 is a Λ_1-map, g_2 a Λ_2-map, and such that

$$\begin{array}{ccc} M \otimes A & \xrightarrow{1 \otimes g_1} & M \otimes A' \\ \Big\downarrow f & & \Big\downarrow f' \\ B & \longrightarrow & B' \end{array}$$

commutes.

The functor $G : \mathrm{mod}(\Gamma) \to \mathcal{J}$ (= category of triples), which gives this equivalence is given by $G(X) = (eX, (1-e)X, f)$, where $e = \begin{pmatrix} 1 & 0 \\ 0 & 0 \end{pmatrix}$. And if we write $M = (1-e)Xe$, $f : M \otimes eX \to (1-e)X$ is given by $f((1-e)\lambda e \otimes em) = (1-e)\lambda em$.

We shall need the special case of algebras of the type $\Gamma = \begin{pmatrix} \Lambda & 0 \\ \mathfrak{R} & \Lambda \end{pmatrix}$, where Λ is a semi-simple algebra, and ${}_\Lambda\mathfrak{R}_\Lambda$ a Λ-bimodule. $\mathfrak{R} = \begin{pmatrix} 0 & 0 \\ \mathfrak{R} & 0 \end{pmatrix}$ is the radical of Γ, and since $\mathfrak{R} \otimes_\Gamma \mathfrak{R} = 0$, it follows by [9] that Γ is hereditary. Viewing the Γ-modules as triples (A,B,f), where A and B are Λ-modules and $f : \mathfrak{R} \otimes A \to B$ a Λ-map, the projective Γ-modules are sums of objects of the form $(A, \mathfrak{R} \otimes A, \mathrm{id})$ and $(0,B,0)$. Let (A,B,f) be an arbitrary object. Since Λ is semi-simple, $\mathrm{Im}\, f \subset B$ splits, so that $B = \mathrm{Im}\, f \oplus B'$. It then follows that $(A,B,f) \cong (A, \mathrm{Im}\, f, f') \oplus (0,B',0)$, where $f' : \mathfrak{R} \otimes A \to \mathrm{Im}\, f$ is the epimorphism induced by f. Hence $\underline{\mathrm{mod}}(\Gamma)$ is the full subcategory of $\mathrm{mod}(\Gamma)$ generated by the indecomposable objects (A,B,f), where f is not a monomorphism.

§2. THE STABLE EQUIVALENCE CLASSES. To a given Artin ring Ω with radical $\mathfrak{R}$ and $\mathfrak{R}^2 = 0$, we consider $\Gamma = \begin{pmatrix} \Omega/\mathfrak{R} & 0 \\ \mathfrak{R} & \Omega/\mathfrak{R} \end{pmatrix} = \begin{pmatrix} \Lambda & 0 \\ \mathfrak{R} & \Lambda \end{pmatrix}$. We consider the additive functor $F : \mathrm{mod}(\Omega) \to \mathrm{mod}(\Gamma)$ given by $F(M) = (M/\mathfrak{R}M, \mathfrak{R}M, f)$, where f is given by

$$\begin{array}{ccc} \mathfrak{R} \otimes M & \xrightarrow{n} & M \\ \Big\downarrow \wr & \nearrow f & \\ \mathfrak{R} \otimes M/\mathfrak{R}M & & \end{array}$$

(n is the natural map). If $g \in Hom_{\Omega}(M,N)$, then $F(g) : F(M) \to F(N)$ is the pair (g_1, g_2) given by the diagram

$$\begin{array}{ccccccccc} 0 & \to & \mathfrak{R}M & \to & M & \to & M/\mathfrak{R}M & \to & 0 \\ & & \downarrow g_1 & & \downarrow g & & \downarrow g_2 & & \\ 0 & \to & \mathfrak{R}N & \to & N & \to & M/\mathfrak{R}N & \to & 0 \end{array}$$

We will use this functor F to prove that Ω and Γ are (projectively) stably equivalent, in the following

THEOREM 2.1. <u>Let Ω be an Artin ring with $\mathfrak{R}^2 = 0$. Then</u> $\Gamma = \begin{pmatrix} \Omega/\mathfrak{R} & 0 \\ \mathfrak{R} & \Omega/\mathfrak{R} \end{pmatrix}$ <u>is an hereditary ring (with $\mathfrak{R}^2 = 0$) such that mod(Ω) and mod(Γ) are equivalent.</u>

PROOF: For M,N Ω-modules, $Hom(FM,FN) = Hom(M,N)/S(M,N)$, where $S(M,N) = \{g : M \to N$ such that $g(M) \subset N\}$. It is also clear that if $(A,B,f) = F(M)$ for some Ω-module M, then f is an epimorphism. In fact, if we denote by $\mathcal{a}$ the full additive subcategory of mod(Γ) consisting of all objects (A,B,f) with f an epimorphism, then "dually" to Chapter 2 in [4], it follows that $F : \text{mod}(\Omega) \to \mathcal{a}$ is a full and dense (i.e. for each C in $\mathcal{a}$, $F(M) \cong C$ for some M) functor. Also, since if $f : M \to M$ is such that $F(f) : F(M) \to F(M)$ is an isomorphism, it follows by the 5-lemma that f is an isomorphism. An additive functor F with these properties is said to give a <u>representation equivalence</u> between mod(Ω) and $\mathcal{a}$ [4]. In particular, it follows that F induces a bijection between the indecomposable objects of the two categories [3].

Denote by F' the restriction of F to $\text{mod}_p(\Omega)$. Let $\mathcal{a}'$ denote the full additive subcategory of $\mathcal{a}$ generated by the objects of the type $F'(M)$. Let $R(M,N) = \{f : M \to N$ such that f factors through a projective Ω-module$\}$. The following lemma will then establish an equivalence of categories between

$\operatorname{mod}(\Omega)$ and $\mathfrak{A}'$.

LEMMA 2.2. <u>If</u> M,N <u>are</u> Ω-<u>modules with no projective summands, then</u> $R(M,N) = S(M,N)$.

PROOF: Assume first that $f \in R(M,N)$, i.e. we have a commutative diagram

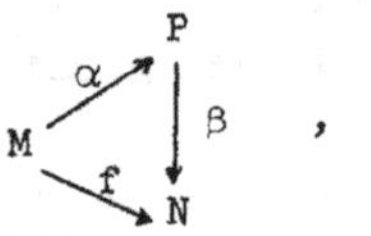

with P a projective Ω-module. Since M has no projective summands, $\alpha(M) \subset \mathfrak{R}P$. Hence $\beta\alpha(M) = f(M) \subset \mathfrak{R}N$, which implies that $f \in S(M,N)$.

Assume conversely that $f \in S(M,N)$, i.e. f factors as follows $M \to M/\mathfrak{R}M \xrightarrow{f'} \mathfrak{R}N \to N$. Let P be a projective cover of N, $P \xrightarrow{g} N \to 0$, g induces an epimorphism $g' : \mathfrak{R}P \to \mathfrak{R}N \to 0$. Since all modules involved are semi-simple, there is an $h : M/\mathfrak{R}M \to \mathfrak{R}P$ which makes the diagram

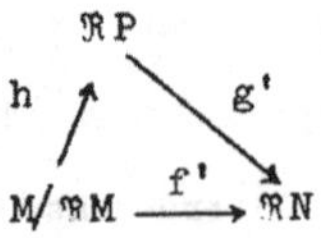

commute. It then follows that f factors through P.

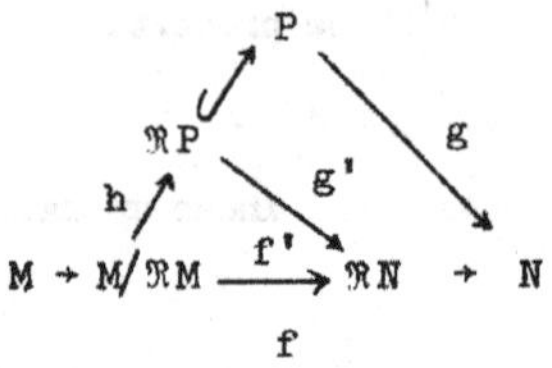

To finish the proof of our theorem, it remains to establish the fact that $\mathfrak{A}'$ and <u>mod(Γ)</u> are equivalent, i.e. that the indecomposable objects in $\mathfrak{A}'$ are exactly the non-projective Γ-modules. The indecomposable projective Γ-modules of type $(0,B,0)$ are already not in $\mathfrak{a}$. We then want to show that an indecomposable Ω-module P is projective if and only if $F(P)$ is of the type $(A, \mathfrak{R} \otimes A, \mathrm{id})$, since these are the remaining indecomposable projective Γ-modules.

LEMMA 2.3. P <u>is a projective</u> Γ-<u>module if and only if</u> $F(P) = (P/\mathfrak{R}P, \mathfrak{R}P, f)$, <u>with</u> f <u>an isomorphism</u>.

PROOF: Assume first that P is a projective Ω-module. Then f is given by

$$\begin{array}{ccc} \mathfrak{R} \otimes P & \xrightarrow{\sim} & P \\ \wr\big\downarrow & \nearrow_{f} & \\ \mathfrak{R} \otimes P/\mathfrak{R}P & & \end{array}$$

And conversely assume that M is not projective, and let P be a projective cover of M, $P \xrightarrow{g} M$. We then have the commutative diagram

$$\begin{array}{ccc} \mathfrak{R} \otimes P & \xrightarrow{\sim} & \mathfrak{R}P \\ \wr\big\downarrow & & \big\downarrow g' \\ \mathfrak{R} \otimes M & \xrightarrow{f} & \mathfrak{R}M \end{array}$$

Since g' is not an isomorphism, f is not.

Before we combine this theorem with the results on hereditary algebras to get our next theorem, we show how to a given Ω to construct a "minimal" hereditary algebra with the same stable category.

As we have seen before, if we decompose an hereditary algebra Γ as $\Gamma = \Gamma' \times S$, where S is semi-simple and Γ' has no semi-simple ring summand, then <u>mod(Γ)</u> and <u>mod(Γ')</u> are equivalent. So from the Γ we constructed as $\Gamma = \begin{pmatrix} \Omega/\mathfrak{R} & 0 \\ \mathfrak{R} & \Omega/\mathfrak{R} \end{pmatrix}$, we want to describe the corresponding Γ'. Let $\Omega_2 = \Omega/\text{left ann}\ \mathfrak{R}$, $\Omega_1 = \Omega/\text{right ann}\ \mathfrak{R}$. Then $\Omega = \Omega_1 \times S_1 = \Omega_2 \times S_2$, for semi-simple rings S_1 and S_2, and $\Gamma = \begin{pmatrix} \Omega_1 & 0 \\ \mathfrak{R} & \Omega_2 \end{pmatrix} \times S_1 \times S_2$. $\Gamma' = \begin{pmatrix} \Omega_1 & 0 \\ \mathfrak{R} & \Omega_1 \end{pmatrix}$

has no semi-simple ring summand, since every division ring which is a ring summand of Ω_1 or Ω_2 has a non-trivial action on $\mathfrak{R}$.

We say that two bimodules ${}_R M_S$ and ${}_{R'} M'_{S'}$ are isomorphic if there are ring isomorphisms $f : R \to R'$ and $g : S \to S'$ and a group isomorphism $h : M \to M'$ such that $h(rm) = f(r)h(m)$, $h(ms) = h(m)g(s)$. We further say that ${}_R M_S$ and ${}_{R'} M'_{S'}$ are <u>related</u> if ${}_{R/(\text{left ann } M)} M_{S/(\text{right ann } M)}$ and ${}_{R'/(\text{left ann } M')} M'_{S'/(\text{left ann } M')}$ are isomorphic.

We are now in the position to state our

THEOREM 2.4. <u>Let Ω and Ω' be Artin algebras with $\mathfrak{R}^2 = \mathfrak{R}'^2 = 0$. Then mod($\Omega$) and mod($\Omega'$) are equivalent, if and only if $\mathfrak{R}$ and $\mathfrak{R}'$ are related in the above sense.</u>

PROOF: If $\mathfrak{R}$ and $\mathfrak{R}'$ are related, the "minimal" hereditary algebras associated with Ω and Ω' will be isomorphic. Hence <u>mod(Ω)</u> and <u>mod(Ω')</u> are equivalent by Theorem 2.2. If conversely <u>mod(Ω)</u> and <u>mod(Ω')</u> are equivalent, the corresponding "minimal" hereditary algebras have equivalent stable categories. By Chapter III they are then isomorphic, which implies that $\mathfrak{R}$ and $\mathfrak{R}'$ are related.

§3. CONSEQUENCES AND EXAMPLES.

COROLLARY 3.1. <u>If we consider Artin algebras with $\mathfrak{R}^2 = 0$, which are trivial extensions of a semi-simple ring summand, then there is only a finite number (up to Morita equivalence) having equivalent stable categories.</u>

PROOF: That Ω has no semi-simple ring summand is equivalent to ${}_\Omega \mathfrak{R}_\Omega$ being a faithful Ω-bimodule, i.e. (left ann $\mathfrak{R}$) $\cap$ (right ann $\mathfrak{R}$) $= 0$. Let as before $\Omega_2 = \Omega/$ left ann $\mathfrak{R}$, $\Omega_1 = \Omega/$ right ann $\mathfrak{R}$. Then $\Omega_1 = \prod_{i=1}^{r} D_i$, $\Omega_2 = \prod_{i=1}^{s} D_i'$, for division rings D_i, D_i'. The other algebras in the same class will be given by different ring automorphisms of Ω_1 and of Ω_2, which will correspond to certain permutations of $\{1, \ldots, r\}$ and $\{1, \ldots, s\}$, so there will be only a finite number of possibilities.

We remark that this finite collection of algebras will all be subrings of the unique hereditary algebra in the class. By our results in Chapter IV there could be no algebras with $\mathfrak{R}^2 \neq 0$ in the same stable equivalence class.

COROLLARY 3.2. _To classify Artin rings with_ $\mathfrak{R}^2 = 0$ _of finite representation type, it is sufficient to consider hereditary rings._

COROLLARY 3.3. _If_ Ω _and_ Ω' _are stably equivalent Artin algebras with_ $\mathfrak{R}^2 = \mathfrak{R}'^2 = 0$, _then_ Ω _is Nakayama if and only if_ Ω' _is._

PROOF: We show that each indecomposable (left) Ω-module has a unique composition series if and only if the same holds for $\Gamma = \begin{pmatrix} \Omega/\mathfrak{R} & 0 \\ \mathfrak{R} & \Omega/\mathfrak{R} \end{pmatrix}$. The indecomposable Γ-modules not of type $F(M)$ are all simple. Further $F(M) = (M/\mathfrak{R}M, \mathfrak{R}M, f)$, with f an epimorphism. Let $\mathfrak{R} = \begin{pmatrix} 0 & 0 \\ \mathfrak{R} & 0 \end{pmatrix}$ be the radical of Γ. Then $F(M) = (0, \mathfrak{R}M, 0)$, and it follows that M is simple if and only if $F(M)$ is. This finishes the proof.

We remark that Corollary 3.3 can be generalized to the case where Ω and Ω' are stably equivalent Artin algebras which are stably equivalent to an hereditary Artin algebra. The proof of this follows from our result of Chapter IV, that the equivalence $H : \overline{\text{mod}(\Omega)} \to \overline{\text{mod}(\Omega')}$, H gives a correspondence between the indecomposable non-simple non-injective projective modules which preserves length and Loewy length. This completes the proof since we have already seen that under these circumstances the indecomposable projective injective modules are uniserial.

EXAMPLE 3.4. Let k be a field, x,y indeterminates, $\Omega = k[x,y]_{(x,y)}/(x,y)^2$. Let $\Omega' = \begin{pmatrix} k & 0 \\ k\oplus k & k \end{pmatrix}$, where the left and right action by k on $k \oplus k$ is given via the diagonal. Then $\underline{\text{mod}(\Omega)}$ and $\underline{\text{mod}(\Omega')}$ are equivalent.

Ω can be written as a trivial extension $\Omega = k \times (k \oplus k)$. Here $\Omega/\mathfrak{R} = k$, and the Γ constructed in our proofs in Ω'.

We point out that Ω and Ω' are different in the way that Ω is commutative and Ω' is not, gl. dim $\Omega = \infty$ and gl. dim $\Omega' = 1$, but still they have equivalent stable categories.

EXAMPLE 3.5. $\Lambda = \begin{pmatrix} k & 0 & 0 \\ k & k & 0 \\ 0 & k & k \end{pmatrix}$ and $\Lambda' = \begin{pmatrix} k & 0 \\ k & k \end{pmatrix} \times \begin{pmatrix} k & 0 \\ k & k \end{pmatrix}$ are stably equivalent.

Here Λ is indecomposable, but Λ' is not.

REFERENCES

[1] Auslander, M., On the dimension of modules and algebras III: Global dimension, Nagoya Math. J. 9, 1955, 67-77.

[2] Auslander, M., Coherent Functors, Proceedings of the Conference on Categorical Algebra, La Jolla, 1965, (Springer-Verlag), 189-231.

[3] Auslander, M., Comments on the Functor Ext, Topology 8 (1969), 151-166.

[4] Auslander, M., Representation dimension of Artin algebras, Queen Mary College Notes, 1971.

[5] Auslander, M., Bridger, M., Stable module theory, Memoirs of the Amer. Math. Society, No. 94, 1969.

[6] Eckmann, B., Hilton, P., Homotopy groups of maps and exact sequences, Comment. Math. Helv. 34, 1960, 271-304.

[7] Gabriel, P., Unzerlegbare Darstellungen I, Manuscripta Mathematico, Vol. 6, Fasc. 1, 1972, 71-103.

[8] Hilton, P., Rees, D., Natural maps of extension functors and a theorem of R. G. Swan, Proc. Camb. Phil. Soc. Math. Phys. Sci. 51, 1961.

[9] Jans, J., Nakayama, T., On the dimension of modules and algebras VII, Algebras with finite dimensional residue-algebras, Nagoya Math. J. 11, 1957, 67-76.

[10] Mitchell, B., On the dimension of objects and categories II, J. of Algebra 9, 1968, 341-368.

[11] Yoshii, T., On algebras of bounded representation type, Osaka Math. J. 8, 1956, 51-105.

SOME PROPERTIES OF TTF-CLASSES*

Goro Azumaya

Indiana University

Since the TTF-theory was initiated by Jans [4], various authors have made interesting studies of the theory, particularly, over semi-perfect rings. The present paper is to establish several basic properties of TTF-classes, which provide a generalization and refinement of their results. It will be shown, among others, that if $(\mathcal{A}, \mathcal{C})$ is a torsion theory over a ring R on the left with TTF-class $\mathcal{C}$ determined by an idempotent ideal I, then $\mathcal{A}$ is a TTF-class or hereditary if and only if I is a direct summand of R or R/I is flat as a right R-module respectively.

Let R be a ring with unit element. A class of (unital) left R-modules is called a torsion class if it is closed under homomorphic images, direct sums, and group extensions, while it is called a torsion-free class if it is closed under submodules, direct products, and group extensions. Let $\mathcal{C}$ be any class of left R-modules. Define $\mathcal{C}^{\ell}$ to be the class of those left R-modules X for which $\mathrm{Hom}_R(X,M) = 0$ for all M in $\mathcal{C}$. Then $\mathcal{C}^{\ell}$ is a torsion class. Similarly, if we define $\mathcal{C}^{r}$ to be the class of those left R-modules Y for which $\mathrm{Hom}_R(M,Y) = 0$ for all M in $\mathcal{C}$, then $\mathcal{C}^{r}$ is a torsion-free class. Conversely, it was shown by Dickson [3] that every torsion class and every torsion-free class is given in this way; more precisely, $\mathcal{C}$ is a torsion class if and only if $(\mathcal{C}^{r})^{\ell} = \mathcal{C}$, while $\mathcal{C}$ is a torsion-free class if and only if $(\mathcal{C}^{\ell})^{r} = \mathcal{C}$. A torsion class $\mathcal{C}$ is called hereditary if it is closed under submodules. A necessary and sufficient condition for this is that the associated torsion-free class $\mathcal{C}^{r}$ is closed under injective envelopes, as was proved also by Dickson. Now, $\mathcal{C}$ is called

*This research was supported by NSF under Grant GP-20434.

a TTF-class if it is both a torsion class and a torsion-free class. This is clearly equivalent to the condition that $\mathcal{C}$ is a torsion-free class which is closed under homomorphic images. It was shown in Jans [4] that there is a one-to-one correspondence between TTF-classes $\mathcal{C}$ and idempotent two-sided ideals I of R in such a way that $\mathcal{C}$ is characterized as the class of those left R-modules which are annihilated by the corresponding I; i.e., $\mathcal{C}$ is actually the class of all left R/I-modules. In this case, it is easy to see that $\mathcal{C}^{\ell}$ is characterized as the class of those left R-modules X which satisfy $IX = X$ and $\mathcal{C}^r$ is the class of those left R-modules Y in which the annihilator of I is 0.

We now start with the following lemma:

LEMMA 1. <u>Let</u> A <u>be a left ideal and</u> B <u>a right ideal of</u> R <u>such that</u> $A + B = R$ <u>and</u> $AB = 0$. <u>Then there exist idempotent elements</u> e, f <u>such that</u> $e + f = 1$, $A = Re$ <u>and</u> $B = fR$, <u>both</u> A <u>and</u> B <u>are two-sided ideals of</u> R, <u>and they are the left annihilator of</u> B <u>and the right annihilator of</u> A <u>in</u> R <u>respectively</u>: $A = l(B)$, $B = r(A)$.

PROOF: Let $e \in A$ and $f \in B$ be such that $e + f = 1$. Then $Re \subset A$. Take any $a \in A$. Then, since $af = 0$, it follows $a = ae + af = ae$. Thus $A = Re$. If we take $a = e$, then we have $e = e^2$. Similarly, $B = fR$ and f is idempotent. Therefore it follows that $r(A) = r(e) = fR = B$ and $l(B) = l(f) = Re = A$. That A is a two-sided ideal follows from $AR = A(A + B) = AA + AB = AA \subset A$. Similarly, B is a two-sided ideal.

PROPOSITION 2. <u>Let</u> A <u>and</u> B <u>be idempotent two-sided ideals of</u> R, <u>and let</u> $\mathcal{A}$ <u>and</u> $\mathcal{B}$ <u>be the TTF-classes corresponding to</u> A <u>and</u> B <u>respectively. Then</u> $\mathcal{A} = \mathcal{B}^{\ell}$, <u>or equivalently</u>, $\mathcal{A}^r = \mathcal{B}$ (<u>or what is the same</u>, $(\mathcal{A}, \mathcal{B})$ <u>forms a torsion theory</u>) <u>if and only if</u> $A + B = R$ <u>and</u> $AB = 0$.

PROOF: Suppose $\mathcal{A} = \mathcal{B}^{\ell}$. Then, since $R/A \in \mathcal{A}$, it follows $B(R/A) = R/A$, which means $A + B = R$. On the other hand, since $B \in \mathcal{A}$ (because of $BB = B$), it follows $AB = 0$. Suppose conversely $A + B = R$ and $AB = 0$. Take any X from $\mathcal{A}$. Then $AX = 0$ and so $X = RX = (A + B)X = AX + BX = BX$. Thus X is in $\mathcal{B}^{\ell}$. Conversely, let X be in $\mathcal{B}^{\ell}$. Then $X = BX$ and therefore $AX = ABX = 0$. Thus X is in $\mathcal{A}$, which shows that $\mathcal{A} = \mathcal{B}^{\ell}$.

THEOREM 3. Let $\mathcal{C}$ be a TTF-class and I the corresponding idempotent two-sided ideal of R. Then $\mathcal{C}^r$ is a TTF-class if and only if $I = Re$ with e an idempotent element of R, while $\mathcal{C}^{\ell}$ is a TTF-class if and only if $I = fR$ with f an idempotent element of R.

PROOF: Suppose $\mathcal{C}^r$ is a TTF-class. Let J be the corresponding idempotent two-sided ideal of R. Then $I + J = R$ and $IJ = 0$ by Proposition 2, and therefore $I = Re$ for some idempotent element e by Lemma 1.

Suppose conversely $I = Re$ with an idempotent element e. Put $J = (1 - e)R$. Then $J = r(I)$, so J is a two-sided ideal and is idempotent since $1 - e$ is idempotent. Since further $eR \subset I$ and $eR + J = eR + (1 - e)R = R$, it follows $I + J = R$. This and the fact that $IJ = 0$ imply, again by Proposition 2, that the TTF-class corresponding to J coincides with $\mathcal{C}^r$.

The other statement for $\mathcal{C}^{\ell}$ can be proved in the similar manner, by using Lemma 1 and Proposition 2 again; this time we have to observe that to $\mathcal{C}^{\ell}$ there should correspond the two-sided ideal $K = l(I)$, which satisfies $K + I = R$.

COROLLARY 4. Let $\mathcal{C}$ be a TTF-class with corresponding idempotent two-sided ideal I. Then both $\mathcal{C}^r$ and $\mathcal{C}^{\ell}$ are TTF-classes if and only if I is a principal ideal generated by a central idempotent element of R, and in this case both the classes coincide: $\mathcal{C}^r = \mathcal{C}^{\ell}$.

PROOF: Let both $\mathcal{C}^r$ and $\mathcal{C}^{\ell}$ be TTF-classes. Then $I = Re = fR$ with idempotent elements e, f by Theorem 3. It follows then $e = fe = f$, and, in addition, for each $a \in R$, we have $ae = fae = fa$ whence $ae = ea$. Thus e is a central idempotent element, which implies $J = r(I) = l(I) = K$; i.e., $\mathcal{C}^r = \mathcal{C}^\ell$.

This corollary is actually the same as Jans [4, Theorem 2.4], and this case is called centrally splitting.

PROPOSITION 5. Let I be a right ideal of R. Then the following conditions are equivalent:

(1) R/I is a flat right R-module.

(2) $I \cap J = IJ$ for all left ideals J of R.

(3) $a \in Ia$ for all $a \in I$.

(4) $IM \cap N = IN$ for every left R-module M and its submodule N.

(The equivalence of (1) and (3) was pointed out to me by Sandomierski.)

PROOF: Since R is flat, the equivalence of (1) and (2) is a particular case of Lambek [6, Proposition 3, p. 133]. Assume (2). Then for any $a \in I$ we have $a \in I \cap Ra = IRa = Ia$. Conversely assume (3). Let J be a left ideal of R. Take any $a \in I \cap J$. Then $a \in Ia \subset IJ$. Thus $I \cap J \subset IJ$ whence $I \cap J = IJ$. Now (2) can be regarded as the special case of (4) where $M = R$ and $N = J$, so that we need only to prove that (3) implies (4). Take any $x \in IM \cap N$ and let $x = a_1u_1 + \dots + a_nu_n$ with $a_i \in I$ and $u_i \in M$. If we assume (3) then by Chase [2, Proposition 2.2] (Villamayor's proposition) there exists $c \in I$ such that $ca_1 = a_1, \dots, ca_n = a_n$. It follows then that $x = ca_1u_1 + \dots + ca_nu_n = cx \in IN$. Thus we have $IM \cap N \subset IN$ whence $IM \cap N = IN$.

THEOREM 6. Let $\mathcal{C}$ be a TTF-class and I the corresponding idempotent two-sided ideal of R. Then $\mathcal{C}^\ell$ is hereditary if and only if R/I is a flat right R-module.

PROOF: $\mathcal{C}^\ell$ consists of those left R-modules M for which $IM = M$, so in particular, the left R-module I is in $\mathcal{C}^\ell$. Suppose now $\mathcal{C}^\ell$ is hereditary. Then for any $a \in I$ the submodule Ra of I is also in $\mathcal{C}^\ell$, and therefore $Ra = IRa = Ia$ whence $a \in Ia$. Thus R/I is flat by Proposition 5. Suppose conversely R/I is flat. Let M be in $\mathcal{C}^\ell$ and N a submodule of M. Then $N = M \cap N = IM \cap N = IN$, again by Proposition 5. This shows that N is also in $\mathcal{C}^\ell$, and thus $\mathcal{C}^\ell$ is hereditary.

COROLLARY 7. <u>Let $\mathcal{C}$ be a TTF-class with corresponding idempotent two-sided ideal I. If $\mathcal{C}^r$ is a TTF-class and $\mathcal{C}^\ell$ is hereditary, then $\mathcal{C}^r = \mathcal{C}^\ell$; i.e., I is a principal ideal generated by a central idempotent element.</u>

PROOF: Since $\mathcal{C}^r$ is a TTF-class by Theorem 3, $I = Re$ with an idempotent element e put $J = (1 - e)R$. Then $J = r(I)$ and so J is a two-sided ideal of R. Since $\mathcal{C}^\ell$ is hereditary, R/I is a flat right R-module by Theorem 6. Therefore $I \cap J = IJ = 0$ by Proposition 5. On the other hand, since $eR \subset I$ and $eR + J = R$, it follows $I + J = R$. These imply that R is a direct sum of I and $J : I \oplus J = R$. Thus I is generated by a central idempotent element.

This corollary is virtually obtained in Bernhardt [1, Theorem 1] and Kurata [5, Theorem 2.7].

THEOREM 8. <u>Let $\mathcal{C}$ be a TTF-class and I the corresponding idempotent two-sided ideal of R. Then $\mathcal{C}^\ell$ is a TTF-class if and only if $\mathcal{C}^\ell$ is hereditary and the left R-module R/I has a projective cover.</u>

PROOF: According to Theorems 3 and 6, it is clear that our theorem is equivalent to the following statement: For a two-sided ideal I of R, $R = fR$ with an idempotent element f if and only if the right R-module R/I is flat and the left R-module R/I has a projective cover. (Observe that if R/I is flat then I is necessarily idempotent because of the condition (3) in Proposition 5.)

Denote by π the natural epimorphism $R \to R/I$. Suppose first that $I = fR$

with an idempotent element f. Put $e = 1 - f$. Then e is idempotent and $R = eR \oplus I$. Therefore R/I ($\cong eR$) is projective, whence flat as a right R-module. Consider next the projective left ideal $Re = l(I)$. Since $Rf \subset I$ and $Re + Rf = R$, it follows $Re + I = R$, or equivalently, $\pi(Re) = R/I$. Let K be any left ideal of R such that $K \subset Re$ and $\pi(K) = R/I$. This however means that $KI = 0$ and $K + I = R$. Then it follows from Lemma 1 that $K = l(I) = Re$. These together imply that Re is a projective cover of the left R-module R/I.

Suppose conversely that the left R-module R/I has a projective cover P. Let $h : P \to R/I$ be the minimal epimorphism. Then, since R is projective, there is a homomorphism $\varphi : R \to P$ such that $h \circ \varphi = \pi$. It follows that the submodule $\varphi(R)$ of P is mapped onto R/I by h and therefore $\varphi(R) = P$. Thus φ is an epimorphism onto the projective module P and so must split, that is, there is a monomorphism $\psi : P \to R$ such that $\varphi \circ \psi = 1_P$, the identity map of P. In particular, the isomorphic image $\psi(P)$ of P is a direct summand, or equivalently, $\psi(P) = Re$ with an idempotent element e. On the other hand, we have $\pi \circ \psi = h \circ \varphi \circ \psi = h \circ 1_P = h$, which means that the diagram

$$\begin{array}{ccccc} P & \xrightarrow{h} & R/I & \longrightarrow & 0 \\ & {\scriptstyle\psi}\searrow & \nearrow{\scriptstyle\pi} & & \\ & & Re & & \end{array}$$

is commutative, thus ψ is an isomorphism. This implies that Re is also a projective cover of R/I with respect to π. Thus it follows that $Re + I = R$ and $Re \cap I = Ie$ is a small submodule of Re. Now, assume in addition, that the right R-module R/I is flat. Take any $a \in I$. Then $ea \in I$, and therefore $Rea = Iea$ by Proposition 5. This means that by means of the epimorphism $Re \to Rea$ (as left R-modules) defined by the mapping $x \to xa$ ($x \in Re$) the submodule Ie of Re is mapped onto Rea, and therefore we have $Re = Ie + (Re \cap l(a))$ because $Re \cap l(a)$ is the kernel of our epimorphism. Since however Ie is small in Re, it follows that $Re = Re \cap l(a)$, i.e., $Rea = 0$. Thus we have proved that $ReI = 0$. We can therefore apply Lemma 1 to Re and I to conclude that $I = fR$ with an idempotent element f (indeed, since $I = r(Re)$ we can choose $f = 1 - e$). This completes the proof of our theorem.

COROLLARY 9. _Let_ R _be semi-perfect and_ $\mathcal{C}$ _a_ TTF-_class. Then_ $\mathcal{C}^{\ell}$ _is a_ TTF-_class whenever it is hereditary._

This follows immediately from Theorem 8 because the semi-perfectness of R implies the existence of the projective cover of the cyclic module R/I.

This corollary is nothing but Rutter [8, Proposition 2]; cf. also [8, Theorem 5].

In connection with the above characterizations of TTF-classes in terms of the corresponding two-sided ideals, we now consider the following aspect of modules. Let I be a two-sided ideal of R and consider the factor ring R/I. Let M be a left R/I-module. Then M can be regarded as a left R-module in the natural manner. It is then easy to see that if M is projective, injective, or flat as an R-module then it is also projective, injective, or flat as an R/I-module respectively. But the converse is not necessarily true, and in fact we have the following characterizations:

PROPOSITION 10. _Let_ I _be a two-sided ideal of_ R. _Then_

(i) _every projective left_ R/I-_module is projective as an_ R-_module if and only if_ $I = Re$ _with an idempotent element_ e _of_ R,

(ii) _every injective left_ R/I-_module is injective as an_ R-_module if and only if_ R/I _is flat as a right_ R-_module,_

(iii) _every flat left_ R/I-_module is flat as an_ R-_module if and only if_ R/I _is flat as a left_ R-_module._

PROOF: (i) "only if" part: Since ${}_{R/I}(R/I)$ is projective, ${}_R(R/I)$ is also projective. Therefore, I is a direct summand of ${}_R R$, that is, I = Re with an idempotent e.

(i) "if" part: Let P be a projective left R/I-module. Then P is a direct summand of a direct sum of (finite or infinite) copies of ${}_{R/I}(R/I)$. But this direct sum is projective when regarded as a left R-module, because I = Re with $e^2 = e$ means that ${}_R(R/I)$ is projective. Thus its direct summand ${}_R P$ is projective.

(ii) "only if" part: Let $a \in I$. Consider Ra, Ia and the factor module Ra/Ia, which are all left R-modules. Let $\varphi: Ra \to Ra/Ia$ be the natural epimorphism. Ra/Ia is annihilated by I and so can be regarded as a left R/I-module. Let Q be the injective envelope of the R/I-module Ra/Ia. Then Q is injective even as an R-module. Therefore φ can be extended to a homomorphism ${}_R R \to {}_R Q$, or, what is the same, there exists $u \in Q$ such that $\varphi(x) = xu$ for all $x \in Ra$. But $Ra \subset I$ and Q is annihilated by I, so $xu = 0$ for all $x \in Ra$. Thus $\varphi = 0$, or Ra = Ia. Since this is true for all $a \in I$, it follows from Proposition 5 that $(R/I)_R$ is flat.

(ii) "if" part: Let Q be an injective left R/I-module. Regarding Q as a left R-module, we denote by E the injective envelope of ${}_R Q$. Let u be an element of E. Let a be an element of I such that $au \in Q$. Then $Iau = 0$. On the other hand, since $(R/I)_R$ is flat, $au \in Iau$ by Proposition 5. Thus we have $au = 0$. This shows that $Iu \cap Q = 0$. But since E is essential over Q, it follows that $Iu = 0$. This is true for every $u \in E$, and thus $IE = 0$, or equivalently, E can be regarded as an R/I-module. Since the R/I-module Q is injective, Q has no essential extension other than itself and therefore we have E = Q. Thus ${}_R Q$ is injective.

(iii) "only if" part: Since ${}_{R/I}(R/I)$ is flat, it follows that ${}_R(R/I)$ is flat.

(iii) "if" part: Let F be a flat left R/I-module. Let M be a right

R-module and N a submodule of M. Since ${}_R(R/I)$ is flat, it follows from (the left-right analogue of) Proposition 5 that $MI \cap N = NI$. This means that N/NI can be regarded as a submodule of the right R-module M/MI in the natural manner. Furthermore, M/MI is annihilated by I and so is looked upon as a right R/I-module. Therefore, by tensoring with the flat R/I-module F (over R/I, or equivalently) over R, we have a monomorphism $(N/NI) \otimes F \to (M/MI) \otimes F$. On the other hand, the exact sequence

$$0 \to MI \to M \to M/MI \to 0$$

yields the following exact sequence

$$MI \otimes F \to M \otimes F \to (M/MI) \otimes F \to 0.$$

But the first term $MI \otimes F = M \otimes IF = M \otimes 0 = 0$, which implies the middle arrow is an isomorphism. Similarly, the homomorphism $N \otimes F \to (N/NI) \otimes F$, induced by the natural epimorphism $N \to N/NI$, is an isomorphism. Thus, by looking at the commutative diagram

$$\begin{array}{ccc} N \otimes F & \longrightarrow & M \otimes F \\ \downarrow & & \downarrow \\ (N/NI) \otimes F & \longrightarrow & (M/MI) \otimes F, \end{array}$$

we can conclude that the top horizontal arrow is a monomorphism, proving the flatness of ${}_RF$.

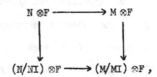

REMARK 1. Theorem 6 can also be derived from Proposition 10, (ii). For, if $\mathcal{C}$ is a torsion-free class, that $\mathcal{C}^{\ell}$ is hereditary is equivalent to the condition that $\mathcal{C}$ is closed under injective envelopes. If, in addition, $\mathcal{C}$ is a TTF-class, then $\mathcal{C}$ consists of those left R-modules which are annihilated by I, the idempotent two-sided ideal corresponding to $\mathcal{C}$, and therefore the latter condition

is equivalent to saying that if a left R-module is annihilated by I then its injective envelope is also annihilated by I; it is however easily seen that the last condition is also equivalent to the condition that every injective left R/I-module is injective as an R-module.

REMARK 2. Proposition 10, (ii) can be regarded as the special case of Morita [7, Lemma 1.3] where $A = R$, $B = R/I$ and $U = R/I$. Also, the "if" part of Proposition 10, (iii) can easily be derived from (the left-right analogue of) the "if" part of (ii) and the Lambek criterion [6, Corollary, p. 131] that ${}_R F$ is flat if and only if its character module F_R^* is injective.

Finally, we would like to prove the following proposition:

PROPOSITION 11. Let P be a left R-module, and let I, T the annihilator ideal, the trace ideal of P respectively. Then the following conditions are equivalent:

(1) ${}_R P$ is projective and $I + T = R$.

(2) ${}_R P$ is projective and ${}_{R/I} P$ is a generator.

(3) ${}_{R/I} P$ is a progenerator and $I = Re$ with an idempotent element $e \in R$.

(4) ${}_R P$ is projective and $T = fR$ with an idempotent element $f \in R$.

PROOF: Since I is the annihilator of ${}_R P$, $I\varphi(P) = \varphi(IP) = 0$ for all $\varphi \in \mathrm{Hom}_R(P,R)$. The trace ideal T of ${}_R P$ is the sum of those $\varphi(P)$'s, so that we have $IT = 0$. Therefore, if we assume that $I + T = R$, then it follows from Lemma 1 that $I = Re$ and $T = fR$ with idempotent elements e and f; observe that both I and T are two-sided ideals. Let $\pi : R \to R/I$ be the natural epimorphism. Then $\pi \circ \varphi \in \mathrm{Hom}_R(P,R/I)$ for all $\varphi \in \mathrm{Hom}_R(P,R)$, and this implies that $\pi(T) = \pi \sum \varphi(P) = \sum \pi \circ \varphi(P)$ is contained in the trace ideal $\overline{T}$ of ${}_{R/I} P$. Therefore, if we assume that $I + T = R$ (i.e., $\pi(T) = R/I$) then it follows $\overline{T} = R/I$; i.e., ${}_{R/I} P$ is a generator. Thus we have proved the validity of the three implications (1) $\Rightarrow$ (2), (1) $\Rightarrow$ (3) and (1) $\Rightarrow$ (4).

(2) $\Rightarrow$ (1) : Since ${}_RP$ is projective, every element of $\mathrm{Hom}_R(P,R/I)$ ($= \mathrm{Hom}_{R/I}(P,R/I)$) is expressed in the form $\pi \circ \varphi$ with an element φ of $\mathrm{Hom}_R(P,R)$, and so we have $\overline{T} = \pi(T)$. However, since ${}_{R/I}P$ is a generator, it follows that $\pi(T) = \overline{T} = R/I$, i.e., $I + T = R$.

(3) $\Rightarrow$ (2) : This is a consequence of the "if" part of Proposition 10, (i).

(4) $\Rightarrow$ (1) : As is well-known, the projectivity of ${}_RP$ implies that $TP = P$. It follows therefore that I is the left annihilator of $T : I = l(T)$. But since $T = fR$, $f^2 = f$, we have that $I = l(f) = R(1 - f)$ and therefore $I + T \supset (1 - f)R + fR = R$. This completes the proof of our proposition.

REFERENCES

[1] R. L. Bernhardt, Splitting hereditary torsion theories over semi-perfect rings, Proc. Amer. Math. Soc. 22 (1969), pp. 681-687.

[2] S. U. Chase, Direct products of modules, Trans. Amer. Math. Soc. 97 (1960), pp. 457-473.

[3] S. E. Dickson, A torsion theory for abelian categories, Trans. Amer. Math. Soc. 121 (1966), pp. 223-235.

[4] J. P. Jans, Some aspects of torsion, Pacific J. of Math. 15 (1965), pp. 1249-1259.

[5] Y. Kurata, On an n-fold torsion theory in the category ${}_R M$, to appear in J. of Algebra.

[6] J. Lambek, Lectures on rings and modules, Blaisdell, 1966.

[7] K. Morita, Localizations in categories of modules I, Math. Z. 114 (1970), pp. 121-144.

[8] E. A. Rutter, Jr., Torsion theories over semi-perfect rings, to appear in Proc. Amer. Math. Soc.

IS $SK_1(Z\pi) = 0$ FOR π A FINITE ABELIAN GROUP

Hyman Bass

Columbia University

My talk was addressed to the problem of the title, discussing background and the relevant avenues of attack, and inviting further attention to it. The lecture was successful in that Roger Alperin, Keith Dennis (who attended the Conference) and Michael Stein, have since solved the problem, using the results of their recent work on K_2 of discrete valuation rings. They prove, in particular, that if π is an elementary p group of order p^n, p an odd prime and $n \geq 1$, then $SK_1(Z\pi)$ is an elementary p group of order p^e, where

$$e = \frac{p^n - 1}{p - 1} - \binom{p + n - 1}{p}$$

Hence, $SK_1(Z\pi) \neq 0$ as soon as $n \geq 3$.

Professor Hyman Bass
Department of Mathematics
Columbia University
New York, New York 10027

STABILITY FOR K_2

R. Keith Dennis

Cornell University

One of the problems considered for the functors K_0 and K_1 was that of "stability." Roughly speaking, one wanted answers to the following two questions:

A. When can $K_0(R)$ be computed in terms of projective modules of bounded rank?

B. When can $K_1(R)$ be computed in terms of matrices of a bounded size?

These questions are discussed in detail in [1] and we will not attempt to answer them here. Rather, we wish to consider an analogous question for the functor K_2:

(*) When can $K_2(R)$ be computed in terms of elementary matrices of a bounded size?

At the present time, a complete answer for the last question is not known. In the following discussion, we give a precise meaning to question (*). A partial answer to this question together with a list of the cases where the results are essentially complete is given below.

Let R be an associative ring with unit. For $n \geq 2$ we denote by $E(n,R)$ the subgroup of the general linear group $GL(n,R)$ generated by the elementary matrices $E_{ij}(r)$. The Steinberg group, $St(n,R)$, is the group with generators $x_{ij}(r)$, where $r \in R$ and i, j are distinct integers between 1 and n, subject to the Steinberg relations

(R1) $\qquad x_{ij}(r)x_{ij}(s) = x_{ij}(r + s)$

(R2) $$[x_{ij}(r), x_{k\ell}(s)] = \begin{cases} 1 & \text{if } i \neq \ell,\ j \neq k \\ x_{i\ell}(rs) & \text{if } i \neq \ell,\ j = k \end{cases}$$

(R3) $$w_{ij}(u)x_{ji}(r)w_{ij}(u)^{-1} = x_{ij}(-uru) \quad \text{for any unit } u \text{ where}$$
$$w_{ij}(u) = x_{ij}(u)x_{ji}(-u^{-1})x_{ij}(u).$$

It should be noted that for $n = 2$, R2 is vacuous and for $n \geq 3$, R3 is a consequence of R1 and R2. As the generators $E_{ij}(r)$ of $E(n,R)$ satisfy relations analogous to R1-R3, there is a surjective homomorphism $St(n,R) \to E(n,R)$ defined by $x_{ij}(r) \to E_{ij}(r)$. We define $K_2(n,R)$ to be the kernel of this homomorphism. For every $n \geq 2$, there is a commutative diagram with exact rows

$$\begin{array}{ccccccccc} 1 & \to & K_2(n,R) & \to & St(n,R) & \to & E(n,R) & \to & 1 \\ & & \downarrow & & \downarrow & & \downarrow & & \\ 1 & \to & K_2(n+1,R) & \to & St(n+1,R) & \to & E(n+1,R) & \to & 1 \end{array}$$

where the vertical maps are defined by sending the generators $x_{ij}(r)$ and $E_{ij}(r)$ in the top row to the element of the same name in the bottom row. Taking direct limits yields the exact sequence

$$1 \to K_2(R) \to St(R) \xrightarrow{\phi} E(R) \to 1$$

which defines $K_2(R)$. $K_2(R)$ is precisely the center of $St(R)$. In fact, the above extension is universal among the central extensions of $E(R)$ and hence there is an isomorphism $H_2(E(R)) \approx K_2(R)$ ([6], [9, §5]). Here $H_2(E(R))$ denotes the second homology group $H_2(E(R); \mathbb{Z})$ where $E(R)$ acts trivially on $\mathbb{Z}$.

If a product of the generators $E_{ij}(r)$ of $E(n,R)$ is trivial, then the corresponding product of the $x_{ij}(r)$'s is an element of $K_2(n,R)$. If this relation among the $E_{ij}(r)$'s is not a consequence of the "universal relations" R1-R3, then the element in $K_2(n,R)$ is non-trivial. This shows that the problem of finding generators for $K_2(n,R)$ is equivalent to the problem of finding a presentation for the group $E(n,R)$. We now give two examples of elements of $K_2(n,R)$:

1. Let α denote a pair of indices ij and let $-\alpha$ denote the reversed pair, ji. For u a unit of the ring R define $w_\alpha(u) = x_\alpha(u)x_{-\alpha}(-u^{-1})x_\alpha(u)$ and $h_\alpha(u) = w_\alpha(u)w_\alpha(-1)$. If $u,v \in R$ are units which commute, then

$$\{u,v\}_\alpha = h_\alpha(uv)h_\alpha(u)^{-1}h_\alpha(v)^{-1}$$

is in $K_2(n,R)$ for $1 \leq i, j \leq n$. If $n \geq 3$, then $\{u,v\}_\alpha$ does not depend on the choice of indices α and is denoted by $\{u,v\}$.

2. If $a,b \in R$ are such that $1 + ab$ is a unit of R, then so is $1 + ba$, and we define

$$H_\alpha(a,b) = x_\alpha(a)x_{-\alpha}(b)x_\alpha(-a(1 + ba)^{-1})x_{-\alpha}(-b(1 + ab)).$$

In case $ab = ba$, the element

$$\langle a,b\rangle_\alpha = H_\alpha(a,b)h_\alpha(1 + ab)^{-1}$$

is in $K_2(n,R)$ for appropriately chosen α. As before, if $n \geq 3$, then $\langle a,b\rangle_\alpha$ does not depend on α and is denoted simply $\langle a,b\rangle$.

Let $m \geq 2$ be a fixed integer We now restate question (*) as follows:

(**) <u>Under what conditions on R will the homomorphisms</u> $K_2(n,R) \to K_2(R)$ <u>be isomorphisms for all</u> $n \geq m$?

In order to answer question B of the introduction, Bass introduced a certain property for rings which also seems to be the relevant property to consider for K_2. An element $(a_1, \ldots, a_m) \in R^m$ is <u>unimodular</u> if there exist $\alpha_1, \ldots, \alpha_m \in R$ such that $\Sigma\alpha_i a_i = 1$. For an integer $m \geq 2$, the ring R satisfies the

<u>stable range condition</u> SR_m if for every unimodular $(a_1, \ldots, a_m) \in R^m$ there exist $b_1, \ldots, b_{m-1} \in R$ such that $(a_1', \ldots, a_{m-1}') \in R^{m-1}$ is unimodular for $a_i' = a_i + b_i a_m$ $(1 \leq i < m)$. In [15] it is shown that the following properties of SR_m hold:

(1) If R satisfies SR_m and S satisfies SR_n, then $R \oplus S$ satisfies SR_k, $k = \max(m,n)$.

(2) If R satisfies SR_m, then so does the opposite ring of R, R^o.

(3) If R satisfies SR_m, then R satisfies SR_n for all $n \geq m$.

(4) If R satisfies SR_m, then the complete ring of $n \times n$ matrices over R satisfies SR_k for $k = 2 - \left[- \frac{m-2}{n} \right]$.

There are numerous examples of rings which satisfy the stability condition SR_m:

(1) Any semi-local ring satisfies SR_2.

(2) Any Dedekind ring satisfies SR_3.

(3) The polynomial ring in one indeterminant over a division ring satisfies SR_3.

(4) (Bass) If R is a finite algebra over a commutative ring A of Krull dimension d, then R satisfies SR_{d+2}.

Proofs of the above statements and additional information can be found in [1, Chapter V], [14] and [15].

If R is a ring satisfying SR_m, then it is easy to show that any matrix $M \in E(n,R)$ for $n \geq m+1$ can be written as a product

$$M = X A L U$$

where X, A, L, U $\in E(n,R)$ satisfy

(i) X has the same first column as the identity matrix,

(ii) A is in the image of $E(m,R) \to E(n,R)$,

(iii) L is lower triangular with 1's on the main diagonal, and

(iv) U is upper triangular with 1's on the main diagonal.

There are obvious subgroups of $St(n,R)$ which correspond to matrices of types X, A, L and U. It is shown in [2] that if the XALU decomposition exists for matrices in $E(m+1,R)$, then the corresponding decomposition exists in $St(n,R)$ for all $n \geq m+1$. Using this normal form in $St(n,R)$, it is possible to prove part (1) of the following theorem; the other three parts follow easily from part (1).

THEOREM 1. Let R satisfy SR_m. Then

(1) The homomorphisms $K_2(n,R) \to K_2(n+1,R)$ are surjective for all $n \geq m+1$;

(2) $K_2(n,R)$ is in the center of $St(n,R)$ for all $n \geq m+2$;

(3) The central extension

$$1 \to K_2(n,R) \to St(n,R) \to E(n,R) \to 1$$

is a universal central extension for all $n \geq \max(m+2,5)$;

(4) $K_2(n,R) \approx H_2(E(n,R))$ for all $n \geq \max(m+2,5)$.

Using a variation on the normal form given above, slightly better results can be obtained in some instances.

THEOREM 2. If R is a commutative ring satisfying SR_3 and for which $SK_1(R)$ is trivial, then $K_2(n,R) \to K_2(n+1,R)$ is surjective for all $n \geq 3$ and $K_2(n,R)$ lies in the center of $St(n,R)$ for all $n \geq 4$.

COROLLARY. *If* $\mathcal{O}$ *is the ring of integers in an algebraic number field, then* $K_2(n,\mathcal{O}) \to K_2(\mathcal{O})$ *is surjective for* $n \geq 3$.

This is the best possible result for surjectivity. It is shown in [3] and [4] that there exist rings of integers $\mathcal{O}$ ($\mathbb{Z}[\sqrt{-17}]$, for example) for which $K_2(2,\mathcal{O}) \to K_2(n,\mathcal{O})$ is not surjective for any $n \geq 3$.

In view of Garland's result [5] that $H_2(E(n,\mathcal{O}))$ is finite for $n \geq 7$, we also obtain the following:

COROLLARY. *If* $\mathcal{O}$ *is the ring of integers in an algebraic number field, then* $K_2(\mathcal{O})$ *is a finite group.*

In the special case where R satisfies SR_2, using yet another variation on the normal form, the following theorem is proved in [2].

THEOREM 3. *Let* R *be a ring which satisfies* SR_2 *and which has the property that any one sided unit of* R *is actually a unit. Then* $K_2(n,R) \to K_2(n+1,R)$ *is surjective for* $n \geq 2$ *and* $K_2(n,R)$ *is central for* $n \geq 3$.

If R is a commutative semi-local ring, then the elements $\langle a,b\rangle$ and $\{u,v\}$ defined earlier generate $K_2(n,R)$ for $n \geq 2$ ([12],[13]). These generators always satisfy certain relations and in some instances they are known to give a presentation for $K_2(n,R)$ for $n \geq 3$ (see [3], [4], [8], [9], [13]).

This brings us back to question (**). Theorems 1, 2 and 3 above only give a partial answer to this question. In analogy with the result for the functor K_1, it is natural to make the following

CONJECTURE. *If* R *satisfies* SR_m, *then the maps* $K_2(n,R) \to K_2(n+1,R)$ *are isomorphisms for* $n \geq m+2$.

It also seems reasonable to ask whether or not $m+2$ can be replaced by $m+1$ in the conjecture. In conjunction with the above, for each integer $n \geq 3$ one would like to have examples of rings R for which the maps $K_2(n,R) \to K_2(n+1,R)$ are not surjective (injective). For $n = 2$, the ring $\mathbb{Z}[\sqrt{-17}]$ provides an example of both properties.

We end this discussion by giving a survey of the cases where the conjecture is known to be true.

I. If $R = \mathbb{Z}$, the ring of rational integers, the maps are surjective for $n \geq 2$ and isomorphisms for $n \geq 3$ [9, §10]. In fact, $K_2(2,\mathbb{Z}) \approx \mathbb{Z}$ and $K_2(n,\mathbb{Z}) \approx \mathbb{Z}_2$ for $n \geq 3$; in all cases, the groups are generated by the symbol $\{-1,-1\}$. In [9] Milnor bases his computation on a lemma of Silvester [11] which in turn generalizes a result of Nielsen [10]. In an earlier version of [9], the computation was based on the results of Magnus [7]. This paper of Magnus contains the first surjective stability theorem for K_2. He used a normal form in $GL(n,\mathbb{Z})$ to obtain a presentation for $GL(n,\mathbb{Z})$ from the one which Nielsen [10] had found for $GL(3,\mathbb{Z})$.

II. If $R = F$ is a field, the maps are surjective for $n \geq 2$ and isomorphisms for $n \geq 3$. This result of Matsumoto [8] (cf. [9, §11, §12]) is obtained by explicitly giving a presentation for the groups $K_2(n,F)$ in terms of their generators, the symbols $\{u,v\}$.

III. If R is a discrete valuation ring or a homomorphic image thereof, the maps are surjective for $n \geq 2$ and isomorphisms for $n \geq 3$. The maps are surjective as all of the groups $K_2(n,R)$ are generated by the symbols $\{u,v\}$. In [4] it is shown that the maps are isomorphisms for $n \geq 3$ by exhibiting defining relations for the groups $K_2(n,R)$.

IV. Using the results of J. R. Silvester, it is shown in [2] that $K_2(n,D) \to K_2(n,D[x])$ is an isomorphism for all division rings D and all integers $n \geq 2$. It can then be shown that $K_2(n,S) \to K_2(n,S[x])$ is an isomorphism for any semi-simple artinian ring S and any $n \geq 3$. The following variation on a lemma originally due to both A. Bak and M. Karoubi can be used to show that $K_2(n,R) \to K_2(n+1,R)$ is an isomorphism for $n \geq 3$ in case R is S or $S[x]$.

LEMMA. Let R be a ring and let $m \geq 3$ be an integer such that

(1) $R[x]$ satisfies SR_m;

(2) The groups $K_2(n,R)$ are abelian for all $n \geq m$;

(3) The homomorphisms $K_2(n,R) \to K_2(n,R[x])$ are isomorphisms for all $n \geq m$.

Then the homomorphisms

$$K_2(n,R) \to K_2(n+1,R)$$

$$K_2(n,R[x]) \to K_2(n+1,R[x])$$

are isomorphisms for all $n \geq m$.

This lemma seems to be of use to prove stability only in the case where R is regular. In particular, property (3) of the Lemma is not valid for the ring $R = \mathbb{Z}/4\mathbb{Z}$, but III above yields a stability result for R. It is known [4] that $K_2(R)$ is cyclic of order two with generator $\{-1,-1\}$. However, the maps from $R[x]$ to R given by $x \to 0$ and $x \to 1$ show that the symbols $\{-1,-1\}$ and $\{-1,1+2x\}$ are independent elements of $K_2(n,R[x])$ and hence $K_2(n,R) \to K_2(n,R[x])$ is not an isomorphism. In fact, in [13] it is shown that $K_2(R[x])$ is an elementary abelian 2-group of countably infinite rank.

REFERENCES

[1] H. Bass, Algebraic K-Theory, Benjamin, New York, 1968.

[2] R. K. Dennis, Surjective stability for K_2, (to appear).

[3] R. K. Dennis and M R. Stein, A new exact sequence for K_2 and some consequences for rings of integers, Bull. Amer Math Soc. 78 (1972), 600-603.

[4] ______________, K_2 of discrete valuation rings, (to appear).

[5] H Garland, A finiteness theorem for K_2 of a number field, Ann. of Math. (2) 94 (1971), 534-548.

[6] M. Kervaire, Multiplicateurs de Schur et K-théorie, pp. 212-225 of Essays on Topology and Related Topics, Memoires dédiés à Georges de Rham (eds. A. Haefliger and R. Narasimhan), Springer-Verlag, Berlin, 1970.

[7] W. Magnus, Über n-dimensionalen Gittertransformationen, Acta Math. 64 (1934), 353-367.

[8] H. Matsumoto, Sur les sous-groupes arithmétiques des groupes semi-simples déployés, Ann. Sci. École Norm. Sup. (4) 2 (1969), 1-62.

[9] J. Milnor, Introduction to Algebraic K-theory, Annals of Math Studies No. 72, Princeton University Press, Princeton, 1971.

[10] J. Nielsen, Die Gruppe der dreidimensionalen Gittertransformationen, Det Kgl. Danske Videnskabernes Selskab. Math-fysiske Meddelelser, V, 12, Kopenhagen (1924), 1-29.

[11] J. R. Silvester, On the K_2 of a free associative algebra, Proc. London Math Soc., (to appear).

[12] M. R. Stein, Surjective stability in dimension 0 for K_2 and related functors, Trans. Amer. Math. Soc., (to appear).

[13] M. R Stein and R. K. Dennis, K_2 of radical ideals and semilocal rings revisited, Lecture Notes in Math., (to appear in the Proceedings of a conference on algebraic K-theory held at the Battelle Institute in Seattle, Washington, August 28 - September 8, 1972).

[14] L. N. Vaserstein, On the stabilization of the general linear group over a ring, Mat. Sbornik 79 (1969), 405-424 (Amer. Math. Soc. Translation v. 8, 383-400.)

[15] ________________, Stable rank of rings and dimensionality of topological spaces, Funkcional. Anal. i Prilozen (2) 5 (1971), 17-27. (Consultants Bureau Translation, 102-110.)

THE THEORY OF RELATIVE GROTHENDIECK RINGS

William H. Gustafson

Indiana University

INTRODUCTION. Let R be a Dedekind domain and G a finite group. By an RG-lattice, we shall mean a left module over the group ring RG which is finitely generated and projective as an R-module. For a variety of topological, arithmetic and group-theoretic reasons, one would like to determine all RG-lattices, up to isomorphism. By induction on R-rank, one may readily show that each lattice is a direct sum of indecomposable lattices (i.e. lattices which cannot be expressed as direct sums of proper sublattices). Hence one should first determine all indecomposable lattices and then give necessary and sufficient conditions that two direct sums of indecomposable lattices be isomorphic. These problems have been solved in only a few special cases. Since we are not at present able to solve them in general, we seek weaker invariants which we find more frequently computable. In this paper, we shall report on a class of such invariants, which take their values in certain rings constructed by K-theoretic means from the category of RG-lattices.

DEFINITION OF THE INVARIANTS. We first note that in the category of RG-lattices, we have a direct sum operation, and also a product: if M, N are RG-lattices, we form $M \otimes_R N$ and impose on it the RG-structure determined by $g \cdot (m \otimes n) = gm \otimes gn$, for $g \in G$, $m \in M$, $n \in N$. These operations inspire us to construct a ring from the RG-lattices. In order to do so, we take note of the facts that we are interested only in isomorphism classes and that additive inverses are thusfar absent. Hence we are led to form the abelian group $a(RG)$ which has as generators the isomorphism classes $[M]$ of RG-lattices, and relations $[M \oplus M'] = [M] + [M']$. Simple calculations show that the product $[M] \cdot [N] =$

$[M \otimes_R N]$ makes $a(RG)$ a commutative ring, the <u>representation ring</u> of RG. The element $[R]$, where G acts trivially on R is an identity element for multiplication. Such rings were first considered by Green [11].

The element $[M] \in a(RG)$ is now an isomorphism invariant of M. How strong an invariant is it? Generally, we have $[M] = [N]$ in $a(RG)$ if and only if $M \oplus X \cong N \oplus X$ for some RG-lattice X. If, in particular, the Krull-Schmidt Theorem holds for RG-lattices (e.g. if R is a complete discrete valuation ring), then $[M] = [N]$ in $a(RG)$ if and only if $M \cong N$. Hence $[M]$ is a very strong invariant, and for this reason, it is nearly as difficult to calculate $a(RG)$ as it is to classify all RG-lattices.

In light of these computational difficulties, we seek to ease our task by allowing more relations in the definition or our ring of invariants. We thus form a ring $K^0(RG)$ in the same fashion as $a(RG)$, except that we say $[M] = [M'] + [M'']$ whenever there is an exact sequence

$$0 \to M' \to M \to M'' \to 0$$

of RG-lattices. This ring, the <u>Grothendieck ring</u> of RG, was introduced and studied extensively by Swan [29,30]. Note that it is a homomorphic image of $a(RG)$.

$K^0(RG)$ is often easy to calculate (at least as an additive group), since it reflects primarily the properties of R-composition series. In the event that R is the ring of integers in an algebraic number field, the abelian group structure of $K^0(RG)$ has, in principle, been determined (see [17]). On the other hand, much of the richness of the category of RG-lattices is often lost in $K^0(RG)$. As an extreme example, consider the case where R is a field of characteristic $p > 0$ and G is a p-group. Then there is only one irreducible RG-module (the trivial one), and the Jordan-Hölder Theorem holds. From these facts, it is easy to deduce that $K^0(RG) \cong \mathbb{Z}$, the isomorphism being given by

$[M] \to \dim_R(M)$. On the other hand, if G is not cyclic, then RG has infinitely many nonisomorphic indecomposable modules [18]. In defense of $K^0(RG)$, it should be noted that when R is a field of characteristic zero, $K^0(RG)$ is isomorphic to the ring of generalized R-characters of G, while if R is a field of finite characteristic, $K^0(RG)$ is isomorphic to the corresponding ring of generalized Brauer characters.

In 1968, T.-Y. Lam and I. Reiner and (independently) A. Dress conceived of a method which makes $a(RG)$ and $K^0(RG)$ special cases of a single definition, and which also supplies a number of rings in between our two examples. The approach is this: fix a subgroup H of G, then form the abelian group with generators the classes $[M]$ and relations $[M] = [M'] + [M'']$ whenever there is an RG-exact sequence

$$0 \to M' \to M \to M'' \to 0$$

which splits when viewed as a sequence of RH-lattices. Multiplying by the same rule as in $a(RG)$, we obtain a ring $a_R(G,H)$, the <u>relative Grothendieck ring of RG relative to</u> H. We observe that $a(RG) = a_R(G,G)$ and $K^0(RG) = a_R(G,1)$ (the latter since RG-lattices are R-projective by definition, so that exact sequences of RG-lattices automatically split over the trivial subgroup). We will survey here some of the properties of $a_R(G,H)$.

<u>MODULAR THEORY</u>. In this section, we denote by F an algebraically closed field of characteristic $p > 0$. Our principal concern will be the case where the order of G is divisible by p, for otherwise, all properties of $a_F(G,H)$ are easily derived from the structure of the semisimple algebra FG.

The principal problem in this area is to prove or disprove the freeness of $a_F(G,H)$ as an abelian group. For, freeness would show that the invariant $[M]$ is expressible uniquely in terms of a fixed set of invariants which would hopefully be canonical in some reasonable sense. One can show that $a(FG)$ is free on $\{[M] \mid M$ is indecomposable$\}$, while $K^{o}(FG)$ is free on $\{[M] \mid M$ is irreducible$\}$; these assertions amount to the Krull-Schmidt and Jordan-Hölder Theorems, respectively. Hence if we define an FG-module to be <u>H-simple</u> if no FG-submodule is an FH-direct summand, we might expect that $a_F(G,H)$ be free on $\{[M] \mid M$ is H-simple$\}$. While it is clear that this set generates $a_F(G,H)$, it is possible to give examples where nontrivial relations exist among the H-simple classes. At present, freeness has not been shown in general, although it has been established in a number of special cases.

Before listing some cases in which freeness is known to hold, let us sketch a situation in which relations hold among the H-simple modules. Our tool will be

THE PUSHOUT LEMMA: <u>Suppose that we are given a commutative diagram</u>

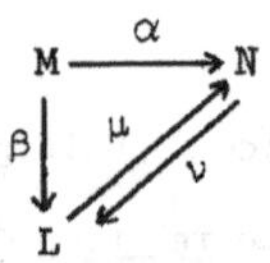

<u>in which</u> M, N, L <u>are</u> FG-<u>modules</u>, α <u>and</u> β <u>are</u> FG-<u>monomorphisms and</u> μ, ν <u>are</u> FH-<u>homomorphisms</u>. <u>Then</u> $[N] - [\text{coker } \alpha] = [L] - [\text{coker } \beta]$ <u>in</u> $a_F(G,H)$.

The lemma takes its name from its proof, which is like that of Schanuel's Lemma.

Now let us assume that F is of characteristic 2, G is cyclic of order 4 with generator y, and $H = \langle y^2 \rangle$ is of order 2. Then there are four indecomposable FG-modules M_1, M_2, M_3, M_4, where $M_n = (y-1)^{4-n}FG$ is an ideal whose F-dimension is n. Among these, M_1, M_3 and M_4 are H-simple, while M_2 lies in the H-split exact sequence

$$0 \to M_1 \to M_2 \to M_2/M_1 \cong M_1 \to 0 .$$

This sequence shows $[M_2] = 2[M_1]$ in $a_F(G,H)$. A bit of matrix calculation produces a diagrom

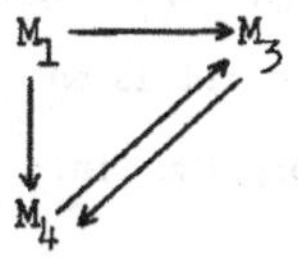

satisfying the hypotheses of the Pushout Lemma. Since $M_3/M_1 \cong M_2$ and $M_4/M_1 \cong M_3$, we derive the relation $[M_4] = 2\cdot([M_3] - [M_1])$ in $a_F(G,H)$. Once this is done, it is simple to show that $a_F(G,H)$ is free on $[M_1]$ and $[M_3]$.

The example above is a special case of

THEOREM (LAM AND REINER [20]): Let G be a finite group, F a field of characteristic p, H a normal cyclic subgroup of G with order p^e. Let s be the number of non-isomorphic principal indecomposable FG-modules. Then $a_F(G,H)$ is free abelian of rank $s\cdot p^e$.

$a_F(G,H)$ is also known to be free in the following cases:

(i) If $K \subset H \triangleleft G$, where $p \nmid [H:K]$ and $[G:H]$ is a power of p, then $a_F(G,K)$ is free (see [25]).

(ii) If G has a normal p-complement, then $a_F(G,H)$ is free for any normal subgroup H of G (see [25]).

(iii) If H is a cyclic subgroup of a p-group G, then $a_F(G,H)$ is free (see [23]).

(iv) If $p = 2$, $|H| = 2$ and $G = \mathrm{Alt}(4)$ or $\mathrm{Sym}(4)$, then $a_F(G,H)$ is free (see [22]).

Let us now turn to some results of a different sort, the reduction theorems. These theorems serve to calculate $a_F(G,H)$ in terms of other (hopefully more accessible) Grothendieck groups.

First, let us assume that K is a normal p-subgroup of G such that $H \cap K = 1$. One would like to excise K; that is, to factor G by K without affecting the relative Grothendieck group $a_F(G,H)$. Thus we put $\overline{G} = G/K$, $\overline{H} = H \cdot K/K \cong H$. Each $F\overline{G}$-module may be regarded as an FG-module on which K acts trivially. This produces a ring homomorphism $\text{inf}: a_F(\overline{G},\overline{H}) \to a_F(G,H)$ (inf for <u>inflation</u>). One then has

EXCISION THEOREM (LAM AND REINER [24]): <u>Suppose that we may write</u> $G = \bigcup_{i=1}^{r} x_i HK$ (<u>disjoint</u>), <u>where</u> $x_1 = 1$ <u>and</u> $\bigcup_{i=1}^{r} Hx_i \subseteq \bigcup_{i=1}^{r} x_i H$. <u>Then</u> $\text{inf}: a_F(\overline{G},\overline{H}) \to a_F(G,H)$ <u>is a ring isomorphism</u>.

COROLLARY: <u>If</u> $G = K \cdot N_G(H)$ <u>or if</u> G <u>is a semidirect produce</u> $K \cdot B$ (<u>where</u> $K \cap B = 1$) <u>and</u> $H \subseteq B$, <u>then inf is a ring isomorphism</u>.

As an application, one can show

PRODUCT ISOMORPHISM THEOREM (LAM AND REINER [24]): <u>Let</u> $H \subseteq E \triangleleft G$, <u>where</u> $G = E \cdot C$ <u>for some subgroup</u> C <u>of the centralizer of</u> E <u>in</u> G. <u>Suppose that</u> $C \cap E$ <u>is a p-group and that</u> F <u>is a splitting field for</u> C <u>and all of its subgroups</u>. <u>Then there is a ring isomorphism</u>

$$a_F(G,H) \cong a_F(G/E,1) \otimes_{\mathbb{Z}} a_F(E,H).$$

This result is especially striking when G is the direct product $E \times C$.

In a slightly different direction, we have the restriction theorems. Given an FG-module M, we may obtain from it an FH-module M_H by restriction of scalars. This restriction process induces a ring homomorphism

$$\text{res}: a_F(G,H) \to a(FH).$$

It is of interest to determine the kernel and image of res. Some progress has been made in this direction in the case where H is a normal subgroup of G.

Let us suppose that that is the situation and that N is an FH-module. For each $g \in G$, we may define a new FH-module N^g, the conjugate of N by g as follows: the underlying F-vector space of N^g is the same as that of N, the element $h \in H$ acts on it as $g^{-1}hg$ acts on N. We say that N is self-conjugate if $N \cong N^g$ for all $g \in G$. Given any FH-module N, the trace of N is the FH-direct sum of a full set of mutually nonisomorphic conjugates of N. It is clearly self-conjugate. If we now denote by T the subgroup of a(FH) generated by all [N] such that N is self-conjugate, we see easily that T is a subring of a(FH) which is additively free on {[trace of N]}, where N ranges over a full set of mutually non-conjugate indecomposable FH-modules. Further, Im(res) $\subseteq$ T. The following theorems relate T and the map res in more detail.

THEOREM (LAM AND REINER [23]): Let F be algebraically closed, and let F^* denote the multiplicative group of nonzero elements of F. Let $H \triangleleft G$ have coset representatives $g_1 = 1, g_2, \ldots, g_r$ and let $g_i g_j = g_k \cdot h_{ij}$, $h_{ij} \in H$. Assume

(i) Each h_{ij} acts trivially on every irreducible FH-module.
(ii) $H^2(E/H, F^*) = 0$, for all subgroups E/H of G/H. Here E/H acts trivially on F^*.

Then the image of res is T.

THEOREM (LAM, REINER AND WIGNER [25]): Let F be algebraically closed of characteristic $p > 0$. Let $H \triangleleft G$ and assume that $[G:H]$ is a power of p. Then res: $a_F(G,H) \to a(FH)$ is a ring isomorphism.

These considerations lead to the following conjecture: Let $H \subseteq E \triangleleft G$, where $[G:E]$ is a power of p. Then the restriction map res: $a_G(G,H) \to a_F(E,H)$ is monic.

It can be shown from the Excision Theorem that it suffices to prove this when G is a split extension of E (see [24, Theorem 4.7]).

These restriction theorems are analogous to some results in the representation theory of Lie groups. Let G be a compact Lie group, T a maximal torus of G and W the Weyl group. Denote by $RU(G)$ the unitary complex representation ring of G and by $RU(T)^W$ the elements of $RU(T)$ fixed under the natural action of W on $RU(T)$. Then restriction from G to T induces an isomorphism $RU(G) \cong RU(T)^W$ (see [2]). Now assume that G acts on a compact space X. Then one can define an equivariant K-theory of bundles on X, obtaining a group $K^*_G(X)$ (see [1]). Matsuda [26] has defined a relative group $K^*_{(G,H)}(X)$ for any closed subgroup H of G and exhibited split exact sequences

$$0 \to K^*_G(X) \to K^*_T(X) \to K^*_{(G,T)}(X) \to 0\,,$$

$$0 \to K^*_G(X) \to K^*_T(X)^W \to K^*_{(G,T)}(X)^W \to 0\,.$$

The pursuit of other analogies with Topological K-theory may well yield additional information on $a_F(G,H)$.

To finish our survey of the modular case, let us turn to the Cartan homomorphism. We first recall that the FG-module M is called <u>(G,H)-projective</u> if each FG-exact sequence

$$0 \to L \to N \to M \to 0$$

which is FH-split is also FG-split. A number of equivalent definitions may be found in [6, Chapter IX]. Denote by $k_F(G,H)$ the subgroup of $a(FG)$ generated by all $[M]$ such that M is (G,H)-projective. We then have a natural mapping $c : k_F(G,H) \to a_F(G,H)$ given by $c[M] = [M]$. This map is the <u>Cartan homomorphism</u>. In the event that $H = \{1\}$, (G,H)-projective means projective in the usual case. Also, the map c then carries the Grothendieck group $P(FG)$ of finitely generated projective FG-modules into the Grothendieck group $K^0(FG)$

of all finitely generated FG-modules. If $P_1, \ldots, P_s$ is a full set of nonisomorphic indecomposable projective FG-modules, and $I_1, \ldots, I_s$ is a full set of nonisomorphic irreducible FG-modules (indexed so that P_j is the projective cover of I_j for each j), then $P(FG)$ is free abelian on $[P_1], \ldots, [P_s]$, while $K^0(FG)$ is free abelain on $[I_1], \ldots, [I_s]$. With respect to these bases, the matrix c is just the classical Cartan matrix [6, p. 593]. One knows that in this classical situation, the determinant of the Cartan matrix is a power of the characteristic p of F. This means of course that the mapping c is monic and that $\operatorname{coker}(c)$ is a finite p-group. Generalizing this, we have

THEOREM (DRESS [7]): <u>The cartan homomorphism</u> $c : k_F(G,H) \to a_F(G,H)$ <u>is a monomorphism whose cokernel is a</u> p-<u>primary group</u>.

As a consequence, one sees that if M and N are (G,H)-projective FG-modules, then $M \cong N$ if and only if $[M] = [N]$ in $a_F(G,H)$.

Perhaps a more important consequence is the following

THEOREM (LAM AND REINER [22]): $a_F(G,H)$ <u>is finitely generated as an abelian group if and only if the</u> p-<u>Sylow subgroups of</u> H <u>are cyclic</u>.

Note that by [18], the condition that the p-Sylow subgroups of H be cyclic is equivalent to the finiteness of representation type of the algebra FH.

Finally, let us show a curiosity that follows from the above theorem of Dress. Suppose that F' is a field containing F. Then we have an obvious mapping $e : a_F(G,H) \to a_{F'}(G,H)$ given by $e[M] = [F' \otimes_F M]$, the <u>ground field extension map</u>. Let us assume that $a_F(G,H)$ is torsion-free. We have a commutative diagram

$$\begin{array}{ccc} k_F(G,H) & \xrightarrow{\ c\ } & a_F(G,H) \\ {\scriptstyle e_1}\downarrow & & \downarrow{\scriptstyle e} \\ k_{F'}(G,H) & \xrightarrow{\ c'\ } & a_{F'}(G,H) \end{array},$$

where e_1 is the obvious map $[M] \to [F' \otimes_F M]$, and c, c' are the appropriate Cartan homomorphisms. From the Krull-Schmidt and Noether-Deuring Theorems, it follows that e_1 is monic. If $x \in \ker(e)$, find a p-power p^n such that $p^n x \in \mathrm{Im}(c)$, say $p^n x = c(y)$. Then $0 = e(p^n x) = ec(y) = c'e_1(y)$. It follows that $y = 0$, since c' and e_1 are monic. As $a_F(G,H)$ is assumed torsion-free, we deduce that $x = 0$. Hence we have

PROPOSITION: <u>In order that relative Grothendieck groups be free, it is necessary that ground field extension always be monic</u>.

<u>INTEGRAL THEORY</u>. In this section we discard the field F and work instead with coefficients from a suitable Dedekind ring R which is not a field. Most of our interest will center around "classical" Dedekind rings of characteristic zero: the rings of integers in number fields and their localizations and completions. We will pursue questions concerning freeness, restriction map, Cartan homomorphism, etc. The principal results will be examples showing that most of the results in the modular case do not carry over to this situation. We have included proofs of some results which were announced without proof in [13].

Our principal examples are in the next two theorems:

THEOREM [13]: <u>Let</u> R <u>be the ring of p-adic integers,</u> G <u>a cyclic group of order</u> p^n <u>and</u> H <u>the unique subgroup of order</u> p. <u>Then</u> $a_R(G,H)$ <u>is free as an abelian group, with rank</u> $n + 2$. <u>As a ring,</u> $a_R(G,H)$ <u>has no nonzero nilpotent elements</u>.

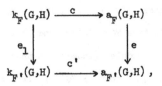

We note that this provides conterexamples to the restriction theorems for

the integral case. We may also note a curiosity: we see above that the rank of $a_R(G,H)$ depends only on n. If we replace R by $\bar{R} = R/\text{rad } R$, the group $a_{\bar{R}}(G,H)$ has rank depending only on p. This points up the fundamental differences between integral and modular representation theory.

THEOREM [13]: _Let_ R _be the ring of_ p-_adic integers_, G _an elementary abelian_ p-_group of order_ p^n _and_ H _any subgroup of order_ p. _Then_ $a_R(G,H)$ _is free as an additive group, with rank_ $(p^n - 1)/(p - 1) + 2$. _As a ring_, $a_R(G,H)$ _has no nonzero nilpotent elements_.

This theorem serves as a counterexample to the integral version of the Excision Theorem.

One may also show that if G is a nonabelian group of order 8, and H is the unique normal subgroup of order 2, then $a_R(G,H) \cong F \oplus V$, where F is free abelian of rank 6, and V is a problematic group of order at most 2. On the basis of such evidence, one may make the following conjecture: If R = p-adic integers, G = p-group and H = normal subgroup of order p, then $a_R(G,H)$ is of rank $e(G) + 1$, where $e(G)$ is the number of simple components in the algebra KG (K = p-adic field). It may well be that $a_R(G,H)$ is free of this rank.

This conjecture would be a direct consequence of the following, which holds in all known cases, both integral and modular:

CONJECTURE: There is an exact "Mayer-Vietoris" sequence

$$0 \to a_R(G,H) \to a_R(G,1) \oplus a_R(H,H) \to a_R(H,1) \to 0 .$$

Here the first map is given by $[M] \to ([M],[M_H])$, while the second is given by $([M],[N]) \to [M_H] - [N]$.

This conjecture would also imply

CONJECTURE: $a_R(G,H)$ is finitely generated as an abelian group if and only if $a(RH)$ is finitely generated.

This latter conjecture is known to be true in the modular case; some evidence for its truth in the integral case will be given below.

Let us now proceed to the Cartan and restriction homomorphisms. These are defined in the same way as in the modular case. Thus let $c : k_R(G,H) \to a_R(G,H)$ denote the Cartan homomorphism. By the same proof as in [21], one may show

THEOREM: Let R be a complete discrete valuation ring with residue field of characteristic $p > 0$. Then c is monic.

COROLLARY: Let R be a Dedekind ring such that the Jordan-Zassenhaus Theorem (see [6]) holds for RG-lattices and the residue fields of R have finite characteristics. Then $\ker c$ is finite.

PROOF: For each prime ideal $P \neq 0$ of R, let R^*_P denote the P-adic completion of R. Then by the theorem, the Cartan homomorphism $c^*_P : k_{R^*_P}(G,H) \to a_{R^*_P}(G,H)$ is monic. We have a commutative diagram

$$\begin{array}{ccc} k_R(G,H) & \xrightarrow{\quad c \quad} & a_R(G,H) \\ \Big\downarrow \varphi & & \Big\downarrow \psi \\ \prod\limits_{P \mid [G:1]} k_{R^*_P}(G,H) & \xrightarrow{\ \Pi c^*_P\ } & \prod\limits_{P \mid [G:1]} a_{R^*_P}(G,H) \end{array}$$

where φ, ψ are the obvious maps. If $c(x) = 0$, then $0 = \psi c(x) = \Pi c^*_P \varphi(x)$, whence $\varphi(x) = 0$ by the theorem. Thus $\ker c \subseteq \ker \varphi = \ker(a(RG) \to \Pi a(R^*_P G))$, and the latter group is finite by [28]. This completes the proof.

COROLLARY: Let R be a discrete valuation ring with finite residue field (R not necessarily complete). Then c is monic.

PROOF: Let R^* denote the completion of R. Then we have a commutative diagram

$$\begin{array}{ccc} k_R(G,H) & \xrightarrow{\ c\ } & a_R(G,H) \\ \varphi\downarrow & & \downarrow \\ k_{R^*}(G,H) & \xrightarrow{\ c^*\ } & a_{R^*}(G,H), \end{array}$$

where $\varphi([M]) = [R^* \otimes_R M]$. Since c^* is monic from the theorem, and φ is monic by [15, Proposition 2.2], it follows that c is monic.

COROLLARY: <u>Let</u> R <u>be a complete discrete valuation ring with residue fields of finite characteristic. Suppose that the Jordan-Zassenhuas Theorem holds for</u> RH-<u>lattices. If</u> $a_R(G,H)$ <u>is finitely generated, then the number</u> $n(RH)$ <u>of indecomposable</u> RH-<u>lattices is finite</u>.

PROOF: Assume that $a_R(G,H)$ is finitely generated, but that $n(RH)$ is infinite. We will obtain a contradiction. By the Jordan-Zassenhaus Theorem, there are indecomposable RH-lattices of arbitrarily large R-rank. Thus for each $i > 0$, we can find an indecomposable RH-lattice M_i of R-rank $> i$. We may also assume $M_i \not\cong M_j$, for $i \neq j$. For each i, we find indecomposable (G,H)-projective RG-lattices $X_1, \ldots, X_k$ such that $M_i^G = RG \otimes_{RH} M_i = \Sigma X_j$. By [6, 63.9], each M_i is an RH-direct summand of $(M_i^G)_H$. Since M_i is indecomposable, it follows from the Krull-Schmidt Theorem for RH-lattices that M_i is an RH-direct summand of $(X_j)_H$ for some j. Thus we have obtained indecomposable (G,H)-projective RG-lattices of arbitrarily large R-rank.

Since the Cartan map $c : k_R(G,H) \to a_R(G,H)$ is monic and $a_R(G,H)$ is finitely generated, $k_R(G,H)$ is finitely generated. But $k_R(G,H)$ is free on $\{[M] \mid M$ is indecomposable and (G,H)-projective$\}$. Thus there are only finitely many indecomposable (G,H)-projectives, so they have bounded R-rank, a contradiction.

COROLLARY: <u>Let</u> K <u>be an algebraic number field which is a splitting field for</u> G , <u>and let</u> R <u>denote the ring of algebraic integers in</u> K . <u>If</u> $a_R(G,H)$ <u>is finitely generated, then</u> $n(RH) < \infty$.

PROOF: Let P be a prime ideal of R . By Lemma 4 of [19] and Proposition 2.5 of [15], the obvious map $a_R(G,H) \to a_{R_P^*}(G,H)$ is surjective. Hence if $a_R(G,H)$ is finitely generated, then so is each $a_{R_P^*}(G,H)$. Therefore $n(R_P^*H) < \infty$ for all P , whence $n(RH) < \infty$ by [19].

In contrast to the "good" results above, let us now show

THEOREM: <u>Let</u> R <u>denote the ring of p-adic integers, let</u> G <u>be a p-group, and let</u> H <u>be a subgroup of</u> G <u>of order</u> p . <u>If</u> $H \neq G$, <u>then the cokernel of the Cartan homomorphism</u> $c : k_R(G,H) \to a_R(G,H)$ <u>has elements of infinite order</u>.

PROOF: Denote by $\bar{R}$ the field R/pR and by K the field of p-adic numbers. The ring RH has three indecomposable lattices A (= R with trivial action of H) , B (= $R[\zeta]$, where ζ is a primitive pth root of unity, and a generator of H acts as multiplication by ζ) and RG . A and B are absolutely indecomposable lattices, i.e. they remain indecomposable under an arbitrary ground ring extension. Since R/pR is perfect, it follows from [27] that A^G and B^G are indecomposable. Also, $RH^G = RG$ is indecomposable by [3]. Therefore $k_R(G,H)$ is free abelian with basis $\{[A^G], [B^G], [RG]\}$. Hence it suffices to show that $a_R(G,H)$ has rank at least four as an abelian group. If G is cyclic of order p^n , $n \geq 2$, then we have seen this already. If G is not cyclic, then it has an elementary abelian quotient G' of order p^2 (see [6, 6.10]). By the Artin Induction Theorem, G' has at least four inequivalent irreducible representations in the rational field $\mathbb{Q}$. Since each irreducible representation of G' is also one of G , we deduce that $K^0(\mathbb{Q}G)$ has rank ≥ 4. By the Noether-Deuring Theorem, $K^0(\mathbb{Q}G)$ is isomorphic to a subgroup of $K^0(KG)$.

Since $a_R(G,H)$ maps onto $K^o(KG)$ (by $[M] \to [K \otimes_R M]$), we see that $a_R(G,H)$ has rank ≥ 4, completing the proof.

As we saw, the restriction map is often well-behaved in the modular case. Let us see that it seldom is in the integral case.

THEOREM: Let R be an integral domain of characteristic zero. Let H be a proper, normal subgroup of a group G. Then the restriction map

$$\text{res} : a_R(G,H) \to a(RH)$$

is not monic.

PROOF: Let M be the RG-lattice determined by the representation of G on the cosets of H. Then M is H-trivial but not G-trivial. Let $x = [M] - [G:H] \cdot [R]$, where G acts trivially on R. Then $\text{res}\, x = 0$, but the image of x in $K^o(RG)$ is nonzero, so x is a nontrivial element of $\ker(\text{res})$.

Finally, let us close by stating the only known theorem on ground ring extension in the integral case

THEOREM (DRESS [9]): Let R be a semilocal Dedekind ring of characteristic zero, and let R^* denote the rad(R)-adic completion of R. Then the completion functor induces a monomorphism

$$a_R(G,H) \to a_{R^*}(G,H) .$$

REFERENCES

[1] M. F. Atiyah, Bott periodicity and the index of elliptic operators, Quart. J. of Math. (Oxford) (2) 19 (1968), 113-140.

[2] R. Bott, Lectures on K(X), W. A. Benjamin, Inc. New York, 1969.

[3] D. B. Coleman, Idemptents in group rings, Proc. Amer. Math. Soc. 17 (1966), 962.

[4] S. B. Conlon, Decompositions induced from the Burnside algebra, J. Algebra 10 (1968), 102-122. Corrections J. Algebra 18 (1971), 608.

[5] __________, Modular representations of $C_2 \times C_2$, J. Austral. Math. Soc. 10 (1969), 363-366.

[6] C. Curtis and I. Reiner, "Representation theory of finite groups and associative algebras," Interscience, New York, 1962.

[7] A. Dress, On relative Grothendieck rings, Bull. Amer. Math. Soc. 75 (1969), 955-958.

[8] __________, On integral and modular relative Grothendieck rings. Multicopied notes of the Summer Open House for Algebraists, Aarhus University (1970), 85-108.

[9] __________, Relative Grothendieckringe über semilokalen Dedekindringen, Surjektivität des Reductionshomomorphismus und ein Theorem von Swan, to appear.

[10] __________, Notes on the theory of representations of finite groups, Part I: The Burnside ring of a finite group and some AGN-applications, Multicopied notes, Universität Bielefeld, 1971.

[11] J. A. Green, The modular representation algebra of a finite group, Ill, J. Math. 6 (1962), 607-619.

[12] __________, Axiomatic representation theory for finite groups, J. Pure and Appl. Algebra 1 (1971), 41-77.

[13] W. Gustafson, Integral relative Grothendieck rings, J. Algebra, 22 (1972), 461-479.

[14] __________, On an induction theorem for relative Grothendieck groups, Proc. Amer. Math. Soc., to appear.

[15] A. Heller, On group representations over a valuation ring, Proc. Nat. Acad. Sci. U.S.A. 47 (1961), 1194-1197.

[16] __________ and I. Reiner, Representations of cyclic groups in rings of integers I, Ann. of Math. (2) 76 (1962), 73-92.

[17] __________ and __________, Grothendieck groups of integral group rings, Ill. J. Math. 9 (1965), 349-360.

[18] D. G. Higman, Indecomposable representations at characteristic p, Duke Math. J. 21 (1954), 377-381.

[19] A. Jones, Groups with a finite number of indecomposable integral representations, Mich. Math. J. 10 (1963), 257-261.

[20] T.-Y. Lam and I. Reiner, Relative Grothendieck groups, J. Algebra 11 (1969), 213-242.

[21] __________ and __________, Reduction theorems for relative Grothendieck rings, Trans. Amer. Math. Soc. 142 (1969), 421-435.

[22] __________ and __________, Finite generation of Grothendieck rings relative to cyclic subgroups, Proc. Amer. Math. Soc. 23 (1969), 481-489.

[23] __________ and __________, Restriction maps on relative Grothendieck rings, J. Algebra 14 (1970), 260-298.

[24] __________ and __________, An excision theorem for Grothendieck rings, Math. Z. 115 (1970), 153-164.

[25] __________, __________ and D. Wigner, Restrictions of representations over fields of characteristic p. Proc. Symposia Pure Math. 21 (1971), 99-106.

[26] T. Matsuda, A relative form of equivariant K-theory, J. Fac. Sci. Shinshu Univ. 5 (1970), 123-139.

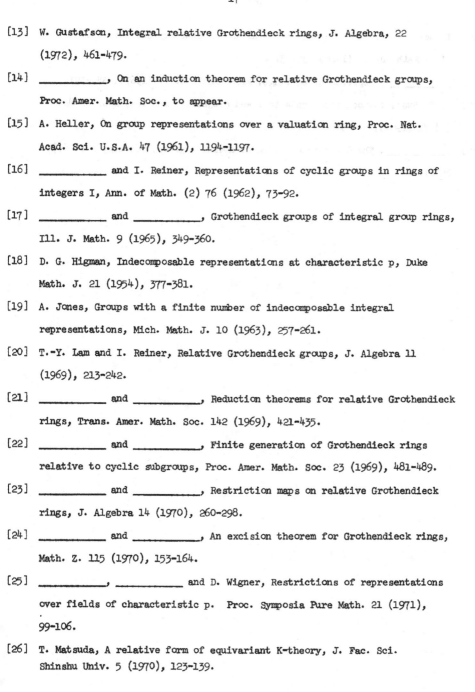

[27] I. Reiner, Relations between integral and modular representations, Mich. Math. J. 13 (1966), 357-372.

[28] __________, Representation rings, Mich. Math. J. 14 (1967), 385-391.

[29] R. G. Swan, Induced representations and projective modules, Ann. of Math. (2) 71 (1960), 552-578.

[30] __________, The Grothendieck ring of a finite group, Topology 2 (1963), 85-110.

THE MORITA CONTEXT AND THE CONSTRUCTION OF QF RINGS

T. A. Hannula

University of Maine at Orono

INTRODUCTION. A left artinian ring R with identity is said to be Quasi-Frobenius, QF for short, when the left R-module ${}_RR$ is injective. It is known [6] that R is QF if and only if either ${}_RR$ or R_R is injective and one of the four chain conditions: left maximum, left minimum, right maximum, or right minimum, holds. In this case R is injective as both a left and right R-module and each of the chain conditions holds. When R is QF,

$$\ell(r(A)) = A \quad \text{and} \quad r(\ell(B)) = B \tag{1.1}$$

for all left ideals A and right ideals B of R where $\ell(X) = \{r \in R \mid rX = 0\}$ and $r(X) = \{r \in R \mid Xr = 0\}$. It follows quickly that the lattice of left ideals of R is dual to the lattice of right ideals of R. We call (1.1) the annihilator duality of the QF ring R. Another useful characterization ([10], Theorem 14.1) of a QF ring is that R is QF if and only if R is artinian and the contravariant functor $\text{Hom}_R(\_,R)$ defines a duality between ${}_R\mathfrak{F}$, the category of finitely generated left R-modules, and $\mathfrak{F}_R$, the category of finitely generated right R-modules. The reader is referred to [4], [5], and [9] for details on the structure of QF rings.

When the ring R can be decomposed into a direct sum of principal indecomposable left R-modules, $R = Rf_1 \oplus \dots \oplus Rf_m$, then a set $\{e_1, \dots, e_n\}$ of primitive orthogonal idempotents of R is said to be basic if Re_i and Re_j are non-isomorphic whenever $i \neq j$ and for each k, $1 \leq k \leq m$, there exists an i such that $Rf_k \cong Re_i$. R is said to be basic when R is a direct sum of mutually non-isomorphic principal indecomposable left ideals. When $\{e_1, \dots, e_n\}$

is a basic set of idempotents of R and $e = e_1 + \dots + e_n$, the subring eRe is a basic ring and is called a basic subring of R. Since $eRe \cong \mathrm{Hom}_R(Re,Re)$, any two basic subrings of R are isomorphic. If B is a basic subring of R, then B and R are Morita equivalent, that is, the category of left B-modules, ${}_B\mathcal{M}$, and the category of left R-modules, ${}_R\mathcal{M}$, are equivalent ([2], [4]). In this case there exists a positive integer t and an idempotent E in B_t, the ring of $t \times t$ matrices over B, such that $R \cong EB_tE$ and $B_tEB_t = B_t$. (See [4], p. 47.) Since the property of being QF is a Morita invariant of a ring, R is QF if and only if its basic subring B is QF. Since the ring R can be recovered from a matrix ring over its basic ring in the above manner, methods for constructing QF rings may be restricted to basic rings.

Originally, Nakayama [11] defined a Quasi-Frobenius ring to be an artinian ring R with 1 such that there is a permutation π on $I = \{e_1, \dots, e_n\}$, a basic set of idempotents of R, such that:

(i) For each i, e_iR has a unique simple right submodule, $\mathrm{Soc}(e_iR)$, and $\mathrm{Soc}(e_iR) \cong \dfrac{\pi(e_i)R}{\pi(e_i)\mathrm{Rad}\ R}$.

(ii) For each i, $R\pi(e_i)$ has a unique simple left submodule, $\mathrm{Soc}(R\pi(e_i))$, and $\mathrm{Soc}(R\pi(e_i)) \cong Re_i/\mathrm{Rad}\ Re_i$.

We call π the Nakayama permutation of the QF ring R. Recently, Fuller [7] established the following result which nicely relates Nakayama's definition of a QF ring, the fact that ${}_RR$ is left injective, and the duality between ${}_R\mathfrak{I}$ and $\mathfrak{I}_R$ induced by R. Here, for a left (right) module M, Soc(M) is the sum of the simple left (right) submodules of M and $T(M) = \dfrac{M}{\mathrm{Rad}\ RM}$ $\left(T(M)) = \dfrac{M}{M\ \mathrm{Rad}\ R}\right)$ where Rad R is the Jacobson radical of R.

FULLER'S THEOREM (THM 3.1, [7]). _If e is an idempotent element in a left artinian ring R, then the following are equivalent:_

(a) Re <u>is injective</u>.

(b) <u>For each</u> e_i <u>in a basic set of idempotents for</u> e <u>there is a primitive idempotent</u> f_i <u>in</u> R <u>such that</u> $Soc(Re_i) \cong T(Rf_i)$ <u>and</u> $Soc(f_iR) \cong T(e_iR)$.

(c) <u>There exists an idempotent</u> f <u>in</u> R <u>such that</u>

(i) $\ell_{fR}(Re) = 0 = r_{Re}(fR)$

(ii) <u>The functors</u> $Hom_{fRf}(\_\_, fRe)$ <u>and</u> $Hom_{eRe}(\_\_, fRe)$ <u>define a duality between the category of finitely generated left</u> fRf-<u>modules and the category of finitely generated right</u> eRe-<u>modules</u>.

<u>Moreover, if</u> Re <u>is injective, then the</u> f_iR <u>of (b) and</u> fR <u>of (c) are also injective</u>.

Fuller's Theorem will play a key role in many of the constructions of QF rings in this paper. As a first consequence, let R be a basic QF ring with Nakayama permutation $\pi = \pi_1 \cdots \pi_k$, a product of disjoint cycles (including 1-cycles) and A_i the orbit of π_i. Then $f_i = \sum_{e_j \in A_i} e_j$ is an idempotent and $f_i = \sum_{e_j \in A_i} \pi^{-1}(e_j)$ since A_i is the orbit of π_i. Since $Soc(Re_i) \cong T(R\pi^{-1}(e_i))$ and $Soc(\pi^{-1}(e_i)R) \cong T(e_iR)$,

$$(1.2) \qquad Soc(Rf_i) = \sum^{\oplus} Soc(Re_i) \cong \sum^{\oplus} T(R\pi^{-1}(e_i)) \cong T(Rf_i)$$

$$(1.3) \qquad Soc(f_iR) = \sum^{\oplus} Soc(\pi^{-1}(e_i)R) \cong \sum^{\oplus} T(e_iR) \cong T(f_iR) .$$

By Fuller's Theorem, (1.2) and (1.3) imply that f_iRf_i defines a duality between ${}_{f_iRf_i}\mathfrak{F}$ and $\mathfrak{F}_{f_iRf_i}$, so f_iRf_i, being an artinian ring, is QF. Moreover, f_iRf_i is basic and the Nakayama permutation of f_iRf_i is a cycle, namely π_i restricted to its orbit A_i. In [8] it was shown that each basic QF ring R could be recovered from the QF subrings f_iRf_i of R and Morita contexts of special form. This construction is given in a slightly modified form in the next section.

Throughout this paper all rings are associative with identity, but subrings need not have the same identity as the overring. All modules are unital, module homomorphisms will be written on the side opposite the scalars, bimodule homomorphisms and ring homomorphisms will be written exponentially.

MORITA CONTEXTS AND THE CONSTRUCTION OF RINGS. Let R and S be rings, ${}_RV_S$ an R-S bimodule, ${}_SW_R$ an S-R bimodule, and $[\ ,\ ] : V \times W \to R$ and $\langle\ ,\ \rangle : W \times V \to S$ be mappings such that

(i) $[\ ,\ ]$ is both left and right R-linear and S-balanced (that is, $[vs,w] = [v,sw]$),

(ii) $\langle\ ,\ \rangle$ is both left and right S-linear and R-balanced,

(iii) $[v,w]v' = v\langle w,v'\rangle$ for all $v,v' \in V$, $w \in W$,

(iv) $\langle w,v\rangle w' = w[v,w'[$ for all $v \in V$, $w,w' \in W$.

The six-tuple $(R, S, V, W, [\ ,\ ], \langle\ ,\ \rangle)$ is called a Morita context. (See [1], [2], [4].) Two Morita contexts $C = (R, S, V, W, [\ ], \langle\ \rangle)$ and $C' = (R', S', V', W', [[\ ]], \langle\langle\ \rangle\rangle)$ are equivalent when there exist ring isomorphisms $\alpha : R \to R'$ and $\beta : S \to S'$ and bijections $\gamma : V \to V'$ and $\delta : W \to W'$ such that $(rvs)\gamma = r^{\alpha}v^{\gamma}s^{\beta}$, $(swr)^{\delta} = s^{\beta}w^{\delta}r^{\alpha}$, $[v,w]^{\alpha} = [[v^{\gamma},w^{\delta}]]$, and $\langle w,v\rangle^{\beta} = \langle\langle w^{\delta},v^{\gamma}\rangle\rangle$ for all $r \in R$, $s \in S$, $v \in V$, and $w \in W$. Observe that the condition of γ insures that the R-S bimodule V' obtained from the ring isomorphisms α and β and the R'-S' bimodule structure of V' by defining $rv's = r^{\alpha}v's^{\beta}$ is isomorphis to V as an R-S bimodule via γ. Similarly, δ is an isomorphism between ${}_SW_R$ and the S-R bimodule obtained from α, β, and the S'-R' bimodule W'.

Given orthogonal idempotents e and f of a ring T such that $e + f = 1$, the Peirce decomposition $T = eTe + fTe + eTf + fTf$ provides a Morita context (eTe, fTf, eTf, fTe, ..., ...) where the mappings eTf $\times$ fTe $\to$ eTe and fTe $\times$ eTf $\to$ fTf are induced by the multiplication of T. Conversely, it was noted in [8] that a Morita context $C = (R, S, V, W, [\ ], \langle\ \rangle)$ yields a ring T in the following manner. Let $T = \left\{ \begin{bmatrix} r & v \\ w & s \end{bmatrix} \middle| r \in R,\ s \in S,\ v \in V,\ w \in W \right\}$, add members of T componentwise, but multiply by

$$(2.1)\qquad \begin{bmatrix} r & v \\ w & s \end{bmatrix} \begin{bmatrix} r' & v' \\ w' & s' \end{bmatrix} = \begin{bmatrix} rr' + [v,w'] & rv' + vs' \\ wr' + sw' & \langle w,v' \rangle + ss' \end{bmatrix}.$$

The ring T is called the ring derived from the Morita context C.

An equivalence $(\alpha, \beta, \gamma, \delta)$ between Morita contexts C and C' yields a ring isomorphism φ, $\begin{bmatrix} r & v \\ w & s \end{bmatrix}^{\varphi} = \begin{bmatrix} r^{\alpha} & v^{\gamma} \\ w^{\delta} & s^{\beta} \end{bmatrix}$, between the rings T and T' derived from the contexts C and C', respectively. When T is the ring derived from the Morita context $C = (R, S, V, W, [\ ], \langle\ \rangle)$ and $e = \begin{bmatrix} 1 & 0 \\ 0 & 0 \end{bmatrix}$, $f = \begin{bmatrix} 0 & 0 \\ 0 & 1 \end{bmatrix}$, then the maps:

$$(2.2)\qquad \begin{aligned} \alpha &: R \to eTe \text{ with } r^{\alpha} = \begin{bmatrix} r & 0 \\ 0 & 0 \end{bmatrix} \\ \beta &: S \to fTf \text{ with } s^{\beta} = \begin{bmatrix} 0 & 0 \\ 0 & s \end{bmatrix} \\ \gamma &: V \to eTf \text{ with } v^{\gamma} = \begin{bmatrix} 0 & v \\ 0 & 0 \end{bmatrix} \\ \delta &: W \to fTe \text{ with } w^{\delta} = \begin{bmatrix} 0 & 0 \\ w & 0 \end{bmatrix} \end{aligned}$$

yield an equivalence $(\alpha, \beta, \gamma, \delta)$ between the Morita context C and the Morita context obtained from the Peirce decomposition of T for the idempotents

e and f. Thus every ring for which the identity is not a primitive idempotent is isomorphic to the derived ring of some Morita context and every Morita context is equivalent to the Morita context obtained from the Peirce decomposition of a ring T with respect to orthogonal idempotents e and $f = 1 - e$ of T.

A local ring is a ring R with identity such that $R/\mathrm{Rad}\, R$ is a division ring. An idempotent e of a ring R is said to be local when eRe is a local ring. If the identity of a ring R has a decomposition $1 = e_1 + \dots + e_n$ such that the e_i are pairwise orthogonal local idempotents, then the number of summands n is uniquely determined and equals the number of irreducible components in a decomposition of the completely reducible $\bar{R}$ (= $R/\mathrm{Rad}\, R$)-module ${}_{\bar{R}}\bar{R}$ ([3] Theorem VII. 1.18). When the identity of R has a decomposition $1 = e_1 + \dots + e_n$ as a sum of pairwise orthogonal local idempotents, n will be called the degree of R, denoted by $\deg R$. Clearly, $\deg R/A \leq \deg R/\mathrm{Rad}\, R = \deg R$ when A is an ideal of R containing $\mathrm{Rad}\, R$. Moreover, in this case $\deg R/A = \deg R$ only when $A = \mathrm{Rad}\, R$.

Let T be the ring derived from the Morita context $C = (R, S, V, W, [\ ], \langle\ \rangle)$. If $\deg R = n$ and $\deg S = m$ with $1 = e_1 + \dots + e_n$ and $1 = f_1 + \dots + f_m$ decompositions of the identities of R and S into a sum of pairwise orthogonal local idempotents, then the identity of T has a decomposition $1 = E_1 + \dots + E_n + F_1 + \dots + F_m$ into pairwise orthogonal local idempotents $E_i = \begin{bmatrix} e_i & 0 \\ 0 & 0 \end{bmatrix}$ and $F_j = \begin{bmatrix} 0 & 0 \\ 0 & f_j \end{bmatrix}$. (Note: $E_iTE_i \cong e_iRe_i$, a local ring since e_i is local.) It follows that $\deg T = \deg R + \deg S$. Note also that in a basic artinian ring R, $\deg R$ is the number of elements in a basic set of idempotents of R.

In a Morita context $(R, S, V, W, [\ ], \langle\ \rangle)$ the mapping $[\ ,\ ]$ is said to be non-degenerate when $w = 0$ whenever $[V,w] = 0$ and $v = 0$ whenever $[v,W] = 0$. The following proposition is a modification of Theorem 3.6 of [8] and gives one method of constructing QF rings by using Morita contexts.

PROPOSITION 1. Let R and S be basic QF rings and $C = (R, S, V, W, [\ ], \langle\ \rangle)$ a Morita context such that

(1) ${}_RV_S$ and ${}_SW_R$ are finitely generated modules with respect to both R and S,

(2) both $[\ ,\ ]$ and $\langle\ ,\ \rangle$ are non-degenerate,

(3) $[V,W] \subseteq \text{Rad } R$ and $\langle W,V\rangle \subseteq \text{Rad } S$.

Then the ring T derived from the Morita context C is a basic QF ring, $\text{Rad } T = \begin{pmatrix} \text{Rad } R & V \\ W & \text{Rad } S \end{pmatrix}$, and the Nakayama permutation π_T of T on the basic set of idempotents $\left\{\begin{bmatrix} e_i & 0 \\ 0 & 0 \end{bmatrix}, \begin{bmatrix} 0 & 0 \\ 0 & f_j \end{bmatrix}\right\}$ is given by

$$\pi_T \begin{bmatrix} e_i & 0 \\ 0 & 0 \end{bmatrix} = \begin{bmatrix} \pi_R(e_i) & 0 \\ 0 & 0 \end{bmatrix}$$

$$\pi_T \begin{bmatrix} 0 & 0 \\ 0 & f_j \end{bmatrix} = \begin{bmatrix} 0 & 0 \\ 0 & \pi_S(f_j) \end{bmatrix} .$$

where π_R (resp. π_S) is the Nakayama permutation on $\{e_i\}$ (resp. $\{f_j\}$) a basic set of idempotents of R (resp. S).

PROOF: Since ${}_RV_S$ and ${}_SW_R$ are finitely generated modules with respect to each of the artinian rings R and S, T is artinian.

Letting $e = \begin{bmatrix} 1 & 0 \\ 0 & 0 \end{bmatrix}$ and $f = \begin{bmatrix} 0 & 0 \\ 0 & 1 \end{bmatrix}$, $eTe \cong R$ and $fTf \cong S$, so eTe and fTf are QF rings. Therefore, eTe defines a duality between ${}_{eTe}\mathfrak{F}$ and $\mathfrak{F}_{eTe}$ and fTf defines a duality between ${}_{fTf}\mathfrak{F}$ and $\mathfrak{F}_{fTf}$. When $x \in \ell_{eT}(Te) =$

$\{x \in eT \mid xTe = 0\}$, $x = \begin{bmatrix} r & v \\ 0 & 0 \end{bmatrix}$ and $\begin{bmatrix} r & v \\ 0 & 0 \end{bmatrix}\begin{bmatrix} R & 0 \\ W & 0 \end{bmatrix} = \begin{bmatrix} rR + [v,W] & 0 \\ 0 & 0 \end{bmatrix} = \begin{bmatrix} 0 & 0 \\ 0 & 0 \end{bmatrix}$.

Thus $r = 0$ and $[v,W] = 0$ whence $v = 0$ by non-degeneracy of [ ]. Similarly, $r_{Te}(eT) = 0$. It now follows from Fuller's Theorem that Te is an injective left T-module. In same manner, $\ell_{fT}(Tf) = 0 = r_{Tf}(fT)$, so Tf is also an injective left T-module. Therefore, T is an artinian ring such that ${}_TT = Te \oplus Tf$ is injective, that is, T is QF.

Since $[V,W] \subseteq \text{Rad } R$ and $\langle W,V\rangle \subseteq \text{Rad } S$, $A = \begin{bmatrix} \text{Rad } R & V \\ W & \text{Rad } S \end{bmatrix}$ is an ideal of T. Moreover, $T/A \cong \frac{R}{\text{Rad } R} \oplus \frac{S}{\text{Rad } S}$ (ring direct sum). Since both $\frac{R}{\text{Rad } R}$ and $\frac{S}{\text{Rad } S}$ are basic when R and S are basic, T/A is a semi-simple basic ring. In particular, $A \supseteq \text{Rad } T$ and $\deg T/A = \deg \frac{R}{\text{Rad } R} + \deg \frac{S}{\text{Rad } S} = \deg R + \deg S = \deg T$, so $A = \text{Rad } T$ and T is basic by the comments on deg T preceding Proposition 1.

To simplify the notation in the following, identify R, S, V, W with the appropriate subsets of T by the maps $\alpha, \beta, \gamma, \delta$ of (2.2). Thus $R = eTe$, $S = fTf$, $V = eTf$, and $W = fTe$. Let U be a simple submodule of Te; let us show that U is a simple left R-submodule of $R = eTe$. First suppose that $eU = 0$. Then $U \subseteq fTe = W$, so $VU \subseteq U \subseteq W$ on one hand, but $VU \subseteq VW \subseteq R$ on the other. Since $W \cap R = 0$, $VU = 0$. But $U \neq 0$, so there exists $u \neq 0$ in U such that $[V,u] = Vu \subseteq VU = 0$ which contradicts the non-degeneracy of [,]. Thus $eU \neq 0$. For any $x \neq 0$ in eU, $eU + fU = U = Tx = Rx + Wx$, since $fx = fex = 0$. But then $Rx = eU$ and eU is a simple left R-module. If $Wx \neq 0$, then $U = TWx = VWx + SWx$. Since $VW = [V,W] \subseteq \text{Rad } R$ and $x \in eU$, a simple left R-module, $VWx = 0$ and $U \subseteq SWx \subseteq fT$, whence $eU \subseteq efT = 0$. Since $eU \neq 0$, $Wx = 0$ and $U = Rx \subseteq eTe = R$. Therefore, each simple T-submodule U of Te is a simple submodule of ${}_RR$. Conversely, if U is a simple left R-submodule of ${}_RR$, then $TU \subseteq Te$ and TU contains a simple T-submodule U' of Te. But then $U' = eU' \subseteq eTU = eTeU = U$, so $U' = U$ and U is a simple T-submodule. Now if $\pi_R(e_i) = e_k$, $e_i \mathrm{Soc}\, Re_k \neq 0$, whence $e_i \mathrm{Soc}\, Te_k \neq 0$ and $\pi_T(e_i) = e_k = \pi_R(e_i)$. Similarly, $\pi_T(f_j) = \pi_S(f_j)$.

REMARK: We can extend the Nakayama permutations π_R and π_S of R and S to $\{e_i\} \cup \{f_j\}$ by letting $\pi_R(f_j) = f_j$, each j, and $\pi_S(e_i) = e_i$, each i. Therefore, we can consider π_T to be the product of the (disjoint) permutations π_R and π_S. In [8] it was shown that any basic QF ring R with Nakayama permutation $\pi = \pi_1 \cdots \pi_k$ with π_i disjoint cycles (with 1-cycles included) could be constructed from basic QF rings R_i with Nakayama permutation π_i and degree R_i = order π_i together with appropriate Morita contexts via the construction in Proposition 1. A basic QF ring of degree n with Nakayama permutation a cycle of order n is called a cyclic QF ring of degree n.

<u>ON THE STRUCTURE OF BASIC QF RINGS</u>. Let R be a basic QF ring of degree n, $\{e_1, \ldots, e_n\}$ a basic set of idempotents, and π the Nakayama permutation of R. In this section we apply Fuller's Theorem to show that the $eRe - \pi(e)R\pi(e)$ bimodule $eR\pi(e)$ is a minimal injective cogenerator with respect to both eRe and $\pi(e)R\pi(e)$. Here ${}_RM$ is a cogenerator (see [2] and [4]) in ${}_R\mathcal{M}$ when M contains a copy of the injective hull of every simple left R-module. In particular, E, the direct sum of the injective hulls of the simple left R-modules, is injective and a direct summand of every cogenerator when R is left noetherian, and in this case E is called a minimal injective cogenerator of ${}_R\mathcal{M}$.

PROPOSITION 2. <u>Let</u> R <u>be a basic</u> QF <u>ring of degree</u> n <u>with Nakayama permutation</u> π <u>and</u> $I = \{e_i \mid 1 \leq i \leq n\}$ <u>a basic set of idempotents of</u> R. <u>Then for each</u> e <u>in</u> I

(i) $eR\pi(e)$ <u>is a minimal injective cogenerator as a left</u> eRe-<u>module and as a right</u> $\pi(e)R\pi(e)$-<u>module</u>,

(ii) $eR\pi(e)$ *induces a duality between* ${}_{eRe}\mathfrak{F}$ *and* $\mathfrak{F}_{\pi(e)R\pi(e)}$.

PROOF: Since $Soc(R\pi e)$ and $Re/Rad\ Re = T(Re)$ are isomorphic as left R-modules, the indecomposable injective left R-module $R\pi(e)$ is paired to the projective right R-module eR in the sense of Fuller [7]. By Lemma 2.2 of [7], the left eRe-module $eR\pi(e)$ is paired to the right eRe-module eRe and $eR\pi(e)$ is the injective hull of $T(eRe) = eRe/Rad(eRe)$. That is, $eR\pi(e)$ is a minimal injective cogenerator for the basic ring eRe. Since $Soc(eR) \cong \pi(e)R/\pi(e)Rad\ R = T(\pi(e)R)$, a left-right symmetric argument shows that $eR\pi(e)$ is a minimal injective cogenerator as a right $\pi(e)R\pi(e)$-module. Thus (i) holds. Since $Soc(R\pi(e)) \cong T(Re)$ and $Soc(eR) \cong T(\pi(e)R)$, (ii) is a direct consequence of Fuller's Theorem quoted in the introduction.

COROLLARY 3. *Let* R *be a basic* QF *ring of degree* n *and* $I = \{e_i \mid 1 \leq i \leq n\}$ *is a basic set of idempotents of* R. *If* eRe *is* QF *for some* $e \in I$, *then* $\pi^i(e)R\pi^i(e) \cong eRe$ *for all positive integers* i.

PROOF: Let eRe be QF, then there is a duality between ${}_{eRe}\mathfrak{F}$ and $\mathfrak{F}_{eRe}$ induced by the regular bimodule eRe. By Proposition 2, there is a duality between ${}_{eRe}\mathfrak{F}$ and $\mathfrak{F}_{\pi(e)R\pi(e)}$, so eRe and $\pi(e)R\pi(e)$ are isomorphic rings by Proposition 4 which follows. In particular, $\pi(e)R\pi(e)$ is QF and the argument may be repeated. Thus for each positive integer i, $eRe \cong \pi^i(e)R\pi^i(e)$.

PROPOSITION 4. *Let* A, B, *and* C *be local noetherian rings such that there exists dualities between* ${}_C\mathfrak{F}$ *and* $\mathfrak{F}_A$ *and between* ${}_C\mathfrak{F}$ *and* $\mathfrak{F}_B$. *Then* A *and* B *are isomorphic rings.*

PROOF: Since ${}_C\mathfrak{F}$ is dual to both $\mathfrak{F}_A$ and $\mathfrak{F}_B$, the categories $\mathfrak{F}_A$ and $\mathfrak{F}_B$ are equivalent. But A and B are noetherian, so $\mathfrak{F}_A$ and $\mathfrak{F}_B$ are abelian categories whose objects are finitely generated, whence the equivalence of $\mathfrak{F}_A$

and $\mathfrak{F}_B$ implies that $\mathfrak{M}_A$ and $\mathfrak{M}_B$ are equivalent, so A and B are Morita equivalent rings. (In [4], replace $\mathfrak{M}_A$ and $\mathfrak{M}_B$ in Proposition 3.1 by $\mathfrak{F}_A$ and $\mathfrak{F}_B$. Modify the proof of (2) $\to$ (3) of Proposition 3.1 by using the Corollary to Theorem 2.6 instead of Theorem 2.6 itself. Finally, apply Theorem 3.3.) Since a Morita class of rings contains at most one basic ring and all local rings are basic, A and B are isomorphic rings.

PROPOSITION 5. Let R be a basic QF ring of degree n, $I = \{e_i \mid 1 \leq i \leq n\}$ a basic set of idempotents of R, and π the Nakayama permutation of R. Then for each e in I such that $\pi(e) \neq e$, $\pi(e)Re \subseteq \text{Rad } Re$ and $eR\pi(e)\pi(e) \subseteq \text{Rad}(eRe)$.

PROOF: Note that $\pi(e)Re\ \text{Soc}(R\pi(e)) \subseteq \text{Soc}(R\pi(e)) \subseteq eR\pi(e)$ and $\pi(e)Re\ \text{Soc}(R\pi(e)) \subseteq \pi(e)ReR\pi(e) \subseteq \pi(e)R\pi(e)$. When $e \neq \pi(e)$, $\pi(e)Re\ \text{Soc}(R\pi(e)) \subseteq eR\pi(e) \cap \pi(e)R\pi(e) = 0$. Since $\text{Soc}(R\pi(e)) \cong Re/\text{Rad } Re$, $\pi(e)Re = \pi(e)ReRe \subseteq \text{Rad } Re$ and $eR\pi(e)\pi(e)Re \subseteq e\text{Rad}Re = \text{Rad}(eRe)$, whenever $\pi(e) \neq e$.

CLASSIFICATION OF CYCLIC QF RINGS OF DEGREE 2 OVER A LOCAL QF RING. Let R be a cyclic QF ring of degree 2 with $\{e,f\}$ a basic set of idempotents of R and the local subring eRe quasi-Frobenius. By Corollary 3, $eRe \cong fRf$ and by Proposition 2, eRf and fRe are minimal injective cogenerators with respect to both eRe and fRf. By constructing a Morita context equivalent to the Peirce decomposition of R with respect to e and f, we show that R is given by a ring of 2×2 matrices over $S = eRe$ with usual addition, but a non-standard product.

Since S is a local QF ring, ${}_SS$ and S_S are minimal injective cogenerators in ${}_S\mathfrak{M}$ and $\mathfrak{M}_S$, respectively. Since S is basic, any minimal injective cogenerator in ${}_S\mathfrak{M}$ (resp. $\mathfrak{M}_S$) is isomorphic to the injective hull of the simple left (resp. right) S-module $S/\text{Rad } S$. Thus there exist isomorphisms of S-modules $\gamma : {}_SeRf \to {}_SS$ and $\delta : fRe_S \to S_S$. Since eRf defines a duality between ${}_S\mathfrak{F}$ and $\mathfrak{F}_{fRf}$, the rings fRf and $\mathrm{Hom}_S(eRf, eRf)$ are isomorphic via the map $t \in fRf \to \varphi_t$ where $x\varphi_t = xt$ for each $x \in eRf$. (See Theorem 6.3 of [10].) Therefore, we have an isomorphism β between the rings fRf and S given by the composition of the isomorphisms $fRf \cong \mathrm{Hom}_S(eRf, eRf) \cong \mathrm{Hom}_S({}_SS, {}_SS) \cong S$ where $\beta : t \to \varphi_t \to \gamma^{-1}\varphi_t\gamma \to (1)\gamma^{-1}\varphi_t\gamma$. Note that

$$(sxt)\gamma = s(xt)\gamma = s(x\varphi_t\gamma) = s(x\gamma\gamma^{-1}\varphi_t\gamma)$$
$$= s((x\gamma)(1)\gamma^{-1}\varphi_t\gamma) = s(x\gamma)t^{\beta}, \text{ so}$$

(4.1) $$(sxt)\gamma = s(x\gamma)t^{\beta}.$$

In the same manner, starting with the isomorphism of right S-modules $\delta : fRE \to S_S$, there exists a ring isomorphism $\beta' : fRf \to S$ such that $(txs)\delta = t^{\beta'}(x\delta)s = (t^{\beta\beta^{-1}})^{\beta'}(x\delta)s = (t^{\beta})^{\rho}(x\delta)s$, where $\rho = \beta^{-1}\beta'$ is an automorphism of the ring S. The automorphism ρ induces an S-S bimodule structure on S via the multiplication $s_1 * s \cdot s_2 = (s_1)^{\rho}ss_2$. This S-S bimodule will be denoted by ${}_{\rho}S$. In particular, we have

(4.2) $$(txs)\delta = t^{\beta} * (x\delta) \cdot s$$

when we consider δ as a bijection from fRe to ${}_{\rho}S$. It follows from (4.1) and (4.2) that $(1, \beta, \gamma, \delta)$ is an equivalence between the Morita context $(eRe, fRf, eRf, fRe, \ldots, \ldots)$ derived from the Peirce decomposition of R and the Morita context $C(R) = (S, S, {}_SS_S, {}_{\rho}S, [\ ], \langle\ \rangle)$ where $[x,y] = (x\gamma^{-1})(y\delta^{-1})$ and $\langle y,x\rangle = ((y\delta^{-1})(x\gamma^{-1}))^{\beta}$. Note that in ${}_{\rho}S$, $s = 1 \cdot s$ on the one hand by considering $1 \in {}_{\rho}S$ as being multiplied on the right by s, but on the other $s = s \cdot 1 = (s^{\rho^{-1}})^{\rho} = s^{\rho^{-1}} * 1$. In particular, for $x \in {}_SS_S$ and $y \in {}_{\rho}S$, $[x,y] = [1 \cdot x,y] = [1, x*y] = [1, x^{\rho}y] = [1,1]x^{\rho}y$, since $[\ ]$ is S-balanced and right S-linear. Also $[x,y] = [x,y \cdot 1] = [x,y^{\rho^{-1}} * 1] = [xy^{\rho^{-1}},1] = [xy^{\rho^{-1}} \cdot 1,1] = xy^{\rho^{-1}}[1,1]$. The following lemmas clarify the structure of the Morita context

$C(R) = (S, S, {}_S S_S, {}_\rho S, [\ ], \langle\ \rangle)$. The first lemma shows that the complication introduced by ρ is necessary whenever ρ is not an inner automorphism of S.

LEMMA 6. <u>The</u> S-S <u>bimodule</u> ${}_\rho S$ <u>is isomorphic to the regular bimodule</u> ${}_S S_S$ <u>if and only if</u> ρ <u>is an inner automorphism</u>.

PROOF: If $\rho : S \to S$ is given by $s^\rho = u^{-1}su$ for some unit u of S, then the map ${}_S S_S \to {}_\rho S$ given by $s \to u^{-1}s$ is an isomorphism of S-S bimodules since $u^{-1}(s_1 s s_2) = u^{-1}s_1 u u^{-1} s s_2 = s_1^\rho u^{-1} s s_2 = s_1 * (u^{-1}s) \cdot s_2$. Conversely, let $\theta : {}_S S_S \to {}_\rho S$ be an isomorphism of S-S bimodules and $u = 1^\theta$. Since $S1 = S$ and $1S = S$, $S * u = S = u \cdot S$ and u is a unit. Moreover, $s^\theta = (s \cdot 1)^\theta = s * 1^\theta = s^\rho u$ and $s^\theta = (1 \cdot s)^\theta = 1^\theta \cdot s = u \cdot s$, whence $s^\rho = usu^{-1}$ and ρ is the inner automorphism determined by u^{-1}.

LEMMA 7. <u>Let</u> R <u>be a cyclic</u> QF <u>ring of degree</u> 2 <u>over the local</u> QF <u>ring</u> S. <u>Then the Morita context</u> $C(R) = (S, S, {}_S S_S, {}_\rho S, [\ ], \langle\ \rangle)$ <u>derived from</u> R <u>has the following properties</u>:

(i) $[1,1] = \langle 1,1 \rangle = u \in \operatorname{Rad} S$

(ii) $u^\rho = u$

(iii) $su = us^\rho$ for all $s \in S$

(iv) $[x,y] = xuy = ux^\rho y$

(v) $\langle y,z \rangle = uxy$.

PROOF: By Proposition 5, $eRffRe \subseteq \operatorname{Rad} eRe$ and $fReeRf \subseteq \operatorname{Rad} fRf$, so $[S, {}_\rho S] = (S\gamma^{-1})({}_\rho S)\delta^{-1} = eRffRe \subseteq \operatorname{Rad} S$ and $\langle {}_\rho S, S \rangle = (({}_\rho S\delta^{-1})(S\gamma^{-1}))^\beta = (fReeRf)^\beta \subseteq (\operatorname{Rad} S)^\beta = \operatorname{Rad} S$. In particular, $u = [1,1] \in \operatorname{Rad} S$. For all $s \in S$, $s[1,1] = [s,1] = [1 \cdot s, 1] = [1, s * 1] = [1, s^\rho] = [1,1]s^\rho$, so $su = us^\rho$. Also $[1,1] = [1,1]1 = 1\langle 1,1 \rangle = \langle 1,1 \rangle$ since $[x,y]x_1 = x\langle y,x_1 \rangle$ for all $x, x_1 \in {}_S S_S$ and $y \in {}_\rho S$. That is, $u = \langle 1,1 \rangle$. Thus $u^\rho = [1,1]^\rho = \langle 1,1 \rangle^\rho = \langle 1,1 \rangle * 1 = 1[1,1] = [1,1] = u$, so $u^\rho = u$. Finally, $[x,y] = [x \cdot 1, 1 \cdot y] =$

$x[1,1]y = xuy = ux^{\rho}y$ and $\langle y,x\rangle = \langle 1\cdot y,x\rangle = \langle 1,yx\rangle = \langle 1,1\rangle yx = uyx$.

Considering the coverse of Lemma 7, we have the following results:

LEMMA 8. Let S be a ring, ρ an automorphism of S and $u \in \text{Rad } S$ such that $u^{\rho} = u$ and $su = s^{\rho}u$ for all $s \in S$. Then $C_{\rho,u} = (S, S, {}_SS_S, {}_{\rho}S, [\ ], \langle\ \rangle)$ is a Morita context where ${}_{\rho}S$ is the S-S bimodule with additive group S and S-S bimodule structure given by $s_1 * s \cdot s_2 = s_1^{\rho} s s_2$, $[\ ,\ ] : {}_SS_S \times {}_{\rho}S \to S$ is given by $[x,y] = xuy$, and $\langle\ ,\ \rangle : {}_{\rho}S \times {}_SS_S \to S$ is given by $\langle y,x\rangle = uys$.

PROOF: Clearly, [ ] is S-linear in both components. Also [ ] is S-balanced, since

$$[xs,y] = xsuy = xus^{\rho}y = [x,s^{\rho}y] = [x,s*y].$$

It is clear that $\langle\ \rangle$ is right S-linear, S-balanced and additive on the left. Moreover,

$$\langle s*y,x\rangle = \langle s^{\rho}y,x\rangle = us^{\rho}yx = suyx = s\langle y,x\rangle.$$

Finally, $x\langle y,x_1\rangle = x(uyx_1) = (xuy)x_1 = [x,y]x_1$ and

$$\begin{aligned} y[x,y_1] &= y(xuy_1) = yux^{\rho}y_1 \\ &= uy^{\rho}x^{\rho}y_1 = u^{\rho}y^{\rho}x^{\rho}y_1 = (uyx)^{\rho}y_1 \\ &= (uyx) * y_1 = \langle y,x\rangle * y_1 . \end{aligned}$$

Thus $C_{\rho,u}$ is a Morita context.

Denote the ring derived from the Morita context $C_{\rho,u}$ by $S_2(\rho,u)$. Then the ring $S_2(\rho,u)$ is the set of 2×2 matrices over S with the usual addition, but multiplication given by

$$\begin{bmatrix} s_1 & s_2 \\ s_3 & s_4 \end{bmatrix}\begin{bmatrix} t_1 & t_2 \\ t_3 & t_4 \end{bmatrix} = \begin{bmatrix} s_1t_1 + us_2^{\rho}t_3 & s_1t_2 + s_2t_4 \\ s_3t_1 + s_4^{\rho}t_3 & us_3t_2 + s_4t_4 \end{bmatrix}.$$

LEMMA 9. Let S be a QF ring and ρ an automorphism of S, then the S-S bimodule ${}_{\rho}S$ defines a duality between ${}_S\mathcal{F}$ and $\mathcal{F}_S$.

PROOF: Since $(r * x)^{\rho^{-1}} = (r^{\rho}x)^{\rho^{-1}} = rx^{\rho^{-1}}$, $\rho^{-1} : {}_{\rho}S \to {}_{S}S$ is an isomorphism of left S-modules. Since S is QF, ${}_{S}S$, hence ${}_{\rho}S$, is a finitely generated injective cogenerator in ${}_{S}\mathcal{M}$. Furthermore, for each left S-endomorphism θ of ${}_{\rho}S$, $\rho\theta\rho^{-1} \in \mathrm{End}_S({}_{S}S)$ so $\rho\theta\rho^{-1} = \varphi_t$ where $t \in S$ and $S\varphi_t = st$ for all $s \in S$. Now $x\theta = x\rho^{-1}\varphi_t\rho = (x^{\rho^{-1}}t)^{\rho} = xt^{\rho}$, so θ is given by right multiplication by t^{ρ} and S is isomorphic to the ring of endomorphisms of the left S-module ${}_{\rho}S$ vis the map $s \to \theta_s$ where $x\theta_s = xs$ for all $x \in {}_{\rho}S$. It now follows that ${}_{\rho}S$ defines a duality between ${}_{S}\mathfrak{F}$ and $\mathfrak{F}_S$ by Morita's Theorem 6.3 in [10].

PROPOSITION 10. <u>If</u> S <u>is a</u> QF <u>ring</u>, ρ <u>an automorphism of</u> S, $u \in \mathrm{Rad}\, S$ <u>such that</u> $u^{\rho} = u$ <u>and</u> $su = us^{\rho}$ <u>for all</u> $s \in S$, <u>then the ring</u> $S_2(\rho, u)$ <u>is</u> QF.

PROOF: Let $e = \begin{pmatrix} 1 & 0 \\ 0 & 0 \end{pmatrix}$, $f = \begin{pmatrix} 0 & 0 \\ 0 & 1 \end{pmatrix}$, and $R = S_2(\rho, u)$. Since S is artinian, R is artinian. To show that R is QF, it is sufficient to show that both Re and Rf are injective left R-modules. Since the Morita context (eRe, fRf, eRf, fRe, .., ..) obtined from the Peirce decomposition of R for the idempotents e and f is equivalent to the Morita context $C_{\rho,u} = (S, S, {}_{S}S_{S}, {}_{\rho}S, [\ ], \langle\ \rangle)$ used to define $R = S_2(\rho, u)$ and both ${}_{S}S_{S}$ and ${}_{\rho}S$ define dualities between ${}_{S}\mathfrak{F}$ and $\mathfrak{F}_S$, eRe and fRf and QF rings, eRf defines a duality between ${}_{eRe}\mathfrak{F}$ and $\mathfrak{F}_{fRf}$, and fRe defines a duality between ${}_{fRf}\mathfrak{F}$ and $\mathfrak{F}_{eRe}$. To show that Re and Rf are injective, it is now sufficient to show

$$\ell_{fR}(Re) = r_{Re}(fR) = 0 \quad \text{and} \quad \ell_{eR}(Rf) = r_{Rf}(eR) = 0$$

by Fuller's Theorem. For $x \in \ell_{fR}(Re)$

$$x = \begin{bmatrix} 0 & 0 \\ s_3 & s_4 \end{bmatrix} \quad \text{and} \quad \begin{bmatrix} 0 & 0 \\ s_3 & s_4 \end{bmatrix}\begin{bmatrix} t_1 & 0 \\ t_3 & 0 \end{bmatrix} = \begin{bmatrix} 0 & 0 \\ 0 & 0 \end{bmatrix}$$

for all t_1 and t_3. Letting $t_3 = 0$ and $t_1 = 1$, $s_3 = 0$. Letting $t_1 = 0$ and $t_3 = 1$, $s_4^{\rho} \cdot 1 = 0$, so $s_4 = 0$. Thus $\ell_{fR}(Re) = 0$. That $r_{Re}(fR) = 0$, $\ell_{eR}(Rf) = 0$, and $r_{Rf}(eR) = 0$ follow in the same manner. Thus both Re and Rf are injective and R is QF.

THEOREM 11. R *is a cyclic* QF *ring of degree* 2 *over a local* QF *ring* S *if and only if* $R \cong S_2(\rho, u)$ *for some automorphism* ρ *of* S, *and* $u \in \text{Rad } S$ *such that* $u^{\rho} = u$ *and* $su = us^{\rho}$ *all* s *in* S.

PROOF: Everything has been established except for the fact that the QF ring $S_2(\rho, u)$ is a basic iing of degree 2 with its Nakayama permutation π a 2-cycle. But $A = \begin{bmatrix} \text{Rad } S & S \\ S & \text{Rad } S \end{bmatrix}$ is an ideal of $S_2(\rho, u)$ and $S_2(\rho, u)/A \cong \frac{S}{\text{Rad } S} \oplus \frac{S}{\text{Rad } S}$ (ring direct sum). As in the proof of Proposition 1, $A = \text{rad } S_2(\rho, u)$ and $S_2(\rho, u)$ is a basic ring of degree 2 with $\left\{ e = \begin{bmatrix} 1 & 0 \\ 0 & 0 \end{bmatrix}, f = \begin{bmatrix} 0 & 0 \\ 0 & 1 \end{bmatrix} \right\}$ a basic set of idempotents. If $f\,\text{Soc}(Re) = 0$, then $fR \cdot \text{Soc}(Re) \subseteq f\,\text{Soc}(Re) = 0$, whence $\text{Soc}(Re) \subseteq r_{Re}(fR)$. But in the proof of Proposition 10 it was shown that $r_{Re}(fR) = 0$, so $\text{Soc}(Re) = 0$ when $f\,\text{Soc}\,Re = 0$. Since $\text{Soc}(Re) \neq 0$, $f\,\text{Soc}(Re) \neq 0$ and $Rf/\text{Rad } Rf \cong \text{Soc}(Re)$. Similarly, $Re/\text{Rad } Re \cong \text{Soc}(Rf)$, so $\pi(f) = e$ and $\pi(e) = f$ and π is a 2-cycle.

COROLLARY 12. *Let* S *be a local* QF *ring and* $u \in \text{Rad } S \cap \text{Center } S$, *then the set of* 2×2 *matrices over* S *form a cyclic* QF *ring* $S_2(u)$ *of degree* 2 *over* S *with the usual addition and multiplication given by*

$$\begin{bmatrix} s_1 & s_2 \\ s_3 & s_4 \end{bmatrix} \begin{bmatrix} t_1 & t_2 \\ t_3 & t_4 \end{bmatrix} = \begin{bmatrix} s_1t_1 + us_2t_3 & s_1t_2 + s_2t_4 \\ s_3t_1 + s_4t_3 & us_3t_2 + s_4t_4 \end{bmatrix}.$$

PROOF: Let ρ be the identity automorphism on S, then $u^{\rho} = u$ and $su = us = us^{\rho}$ for all $s \in S$. It is now easy to check that

$$S_2(u) = S_2(\rho, u),$$

a cyclic QF ring of degree 2 over S.

REFERENCES

[1] S. A. Amitsur, Rings of Quotients and Morita Context, J. Algebra, 17 (1971), 273-298.

[2] H. Bass, Algebraic K-Theory, W. A. Benjamin, Inc., New York, N.Y., 1968.

[3] Ernst-August Behrens, Ring Theory, Academic Press, New York, N.Y., 1972.

[4] P. M. Cohn, Morita Equivalence and Duality, Queen Mary College Lecture Notes, 1968.

[5] C. W. Curits and I. Reiner, Representation Theory of Finite Groups and Associative Algebras, Interscience, New York, N.Y., 1962.

[6] Carl Faith, Rings with Ascending Condition on Annihilators, Nagoya Math J., 27 (1966), 179-191.

[7] Kent Fuller, On Indecomposable Injectives over Artinian Rings, Pacific J. Math., 29 (1969), 115-135.

[8] T. A. Hannula, On the Construction of Quasi-Frobenius Rings, J. Algebra, to appear.

[9] J. P. Jans, Rings and Homology, Holt, New York, N.Y., 1964.

[10] K. Morita, Duality for Modules and its Applications to the Theory of Rings with Minimum Condition, Science Reports of the Tokyo Kyoiku Daigaku, Sect. A. 6 (1958), 83-112.

[11] T. Nakayama, On Frobeniusean Algebras II, Ann. of Math., 42 (1941), 1-21.

ON STEINBERG GROUPS

Stan Klasa

McGill University

In (4) J. Milnor defined the groups $St_n(R)$ and homomorphisms $\varphi_n : St_n(R) \to E_nR$ for an arbitrary associative unitary ring R.

Let n be a positive integer greater than 2. Let us recall that $St_n(R)$ is defined by generators x_{ij}^{λ} (denoted also for typographical reasons by $x_{ij}(\lambda)$), where i, j are positive integers less than or equal to n, $i \neq j$, and $\lambda \in R$ and subject to the relations

$$\text{(i)} \quad x_{ij}(\lambda) \cdot x_{ij}(\mu) = x_{ij}(\lambda + \mu),$$

$$\text{(ii)} \quad [x_{ij}(\lambda), x_{kl}(\mu)] = 1,$$

$$\text{(iii)} \quad [x_{ij}(\lambda), x_{jl}(\mu)] = x_{il}(\lambda\mu)$$

$$(1 \leq i, j, k, l \leq n, \ i \neq l, \ j \neq k; \ \lambda, \mu \in R;$$
$$[a,b] = aba^{-1}b^{-1}).$$

The group $E_n(R)$ is the subgroup of $GL_n(R)$ generated by <u>elementary</u> matrices $E_{ij}(\lambda)$ of the form $I_n + \lambda e_{ij}$, $i \neq j$, where (e_{ij}) is the canonical basis of the module of $n \times n$ matrices and $\varphi_n(x_{ij}(\lambda)) = E_{ij}(\lambda)$.

Similarly one defines the "infinite" groups $St(R)$ and $E(R)$ as direct limits of $St_n(R)$ and $E_n(R)$, (cf. (4) and H. Bass (1)).

All these groups are perfect (a group G is perfect if $G = [G,G]$). In addition, in the infinite case, we have $E(R) = [GL(R),GL(R)]$ and the homomorphism $\varphi : St(R) \to E(R)$ is central. The last statement means that the kernel $Ker(\varphi)$ is contained in the center of $St(R)$, (or equivalently $[Ker\ \varphi, St(R)] = 1$). It is precisely this group $Ker(\varphi)$ that was chosen by

J. Milnor as a definition of his famous $K_2(R)$.

The Schur cover $Sch_n(R)$ of $E_n(R)$ is uniquely defined as maximal central extension of $E_n(R)$:

(a) $1 \to K \overset{\iota}{\to} Sch_n(R) \overset{\varepsilon}{\to} E_n(R) \to 1$ (exact)

(b) $[K, Sch_n(R)] = 1$

(c) $[Sch_n(R), Sch_n(R)] = Sch_n(R)$

such that

$$1 \to K' \overset{\iota'}{\to} G \overset{\varepsilon'}{\to} E_n(R) \to 1 \quad \text{(exact)}$$

$$[\iota' K', G] = 1$$

implies that there is a uniquely determined homomorphism α from $Sch_n(R)$ to G with commutative diagram:

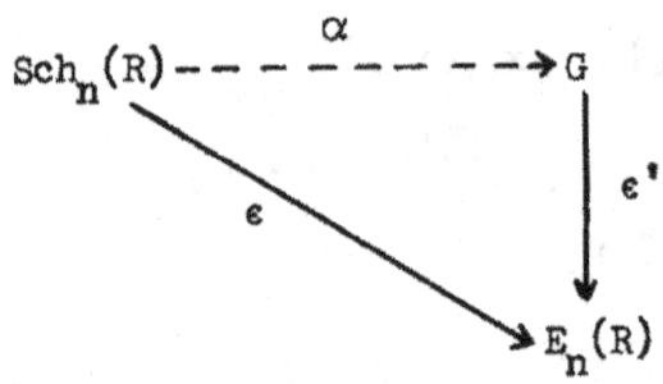

(see Schur (5)).

The Steinberg group $St_n(R)$ is not a central extension of $E_n(R)$ in general. Nevertheless we shall show

THEOREM 1. For any central extension

$$1 \to K \to G \overset{\psi}{\to} E_n(R) \to 1, \quad \text{with } n \geq 5,$$

the group G is a quotient of the Steinberg group $St_n(R)$. More precisely there is a unique group epimorphism

$$h : St_n(R) \to G$$

<u>making the following diagram</u>

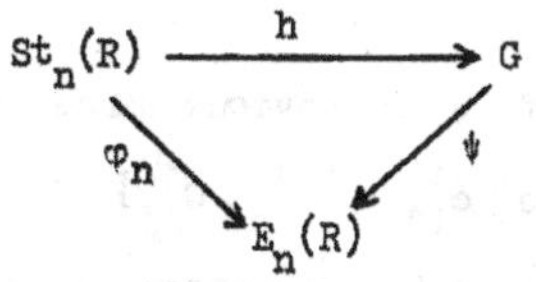

<u>commutative</u>.

COROLLARY 1. <u>If</u> φ_n <u>is central, then</u> $St_n(R)$ <u>is isomorphic to the universal central extension</u> $Sch_n(R)$ <u>of</u> $E_n(R)$.

COROLLARY 2. (Kervaire-Milnor). <u>The groups</u> $St(R)$ <u>and</u> $Sch(R)$ <u>are canonically isomorphic</u>.

PROOF OF COROLLARY 1 : Take $G = Sch_n(R)$; we know by Theorem 1 the existence of a unique morphism h :

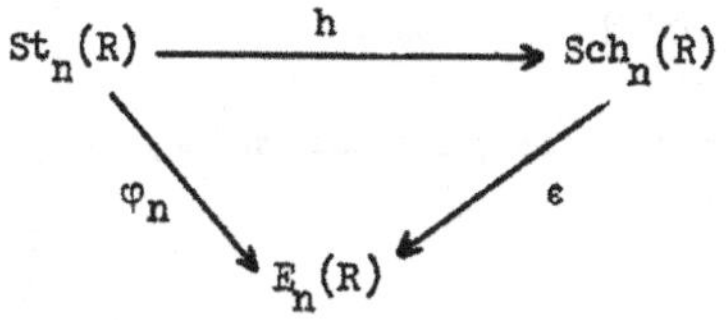

But if $St_n(R)$ happens to be a central extension of $E_n(R)$, the universality of $Sch_n(R)$ provides another epimorphism

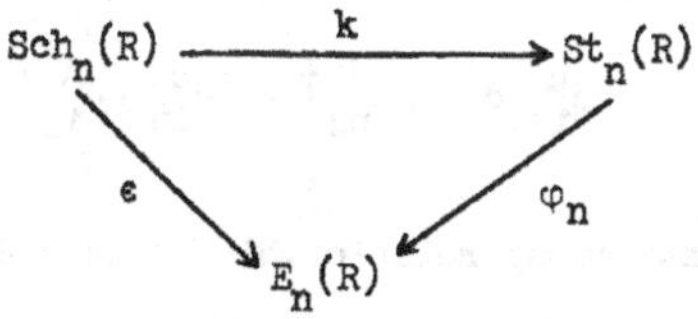

and necessarily $h \cdot k$ and $k \cdot h$ are identities.

PROOF OF COROLLARY 2. Theorem 1 is valid for any $n \geq 5$, and therefore is valid also for limit groups St(R) and E(R). By Milnor's theorem, we know that St(R) is a central extension of E(R). By using the same argument as above, we deduce that St(R) is the universal central extension of E(R).

PROOF OF THEROEM 1. The uniqueness of h is obvious since $St_n(R)$ is perfect. To show the existence of h, we choose $c_{ij}^{\lambda} = [c_{in}^{\lambda}, c_{nj}^{1}]$ for $n \notin \{i.j\}$, in cosets $C_{ij}^{\lambda} = \varphi_n^{-1}(E_{ij}(\lambda))$ corresponding to generators of $E_n(R)$ and define $h(x_{ij}(\lambda)) = c_{ij}^{\lambda}$. All we have to check is the independence of c_{ij}^{λ} on the choice of n and Steinberg relations. The following verification relies on only one "commutator identity":

$$[ab,c] = [a,[bc]] \cdot [b,c] \cdot [a,c].$$

(a) $[c_{ij}^{\lambda}, c_{kl}^{\mu}] = 1$ if $j \neq k$, $i \neq l$.

Take $a \in C_{ij}^{\lambda}$, $c \in C_{kn}^{\mu}$, $d \in C_{nl}^{1}$, $n \notin \{i,j,k,l\}$, $b = [c,d]$.

Than $b^a = [c,d]^a = [c^a,d^a] = [c,d] = b$.

(b) $[c_{ik}^{\lambda}, c_{kj}^{\mu}] = [c_{in}^{\lambda\mu}, c_{nj}^{1}]$.

Taking $\mu = 1$ we get the independence on n, and then the Steinberg relation (iii).

PROOF OF (b): Take $a \in C_{ik}^{\lambda}$, $c \in C_{kn}^{\mu}$, $d \in C_{nj}^{1}$, $b = [c,d] \in C_{kj}^{\mu}$, $n \neq k$,

$$b^a = [c^a,d^a] = [[a,c]c,d] = [[a,c],b] \cdot b \cdot [[a,c],d] = [c_{in}^{\lambda\mu}, c_{nj}^{1}] \cdot b.$$

(c) $c_{ij}^{\lambda} \cdot c_{ij}^{\mu} = c_{ij}^{\lambda+\mu}$.

$$c_{ij}^{\lambda+\mu} = [c_{in}^{\mu+\lambda}, c_{nj}^{1}] = [c_{ij}^{\mu} c_{in}^{\lambda}, c_{nj}^{1}] = [c_{in}^{\mu}, c_{ij}^{\lambda}]\, c_{ij}^{\lambda}\, c_{ij}^{\mu} = c_{ij}^{\lambda}\, c_{ij}^{\mu}.$$

A natrual question is to ask about relationship between Steinberg groups as well as the group $K_2(R)$ of a given ring R and those of some rings produced from R (like rings of matrices for example). Also it is important to know when $K_2(R)$ may be computed using "finite" Steinberg groups.

For any unit ring R let A be the ring $M_n(R)$ of $n \times n$ matrices with entries in R. We shall see that for any integer $p \geq 3$ there is an epimorphism between finite Steinberg groups

$$\tilde{\alpha} : St_p(A) \twoheadrightarrow St_{np}(R)$$

besides the fact that "infinite" groups $St(A)$ and $St(R)$ are isomorphic.

First let us remark that for any positive integer p, we have an obvious isomorphism

$$\underline{\alpha} : GL_p(A) \to GL_{np}(R) .$$

This gives rise, after passing to the limit, to the following isomorphisms:

$$\begin{aligned}
GL(A) &= \varinjlim_p GL_p(A) &\cong \varinjlim_p GL_{np}(R) &= GL(R) \\
E(A) &= GL'(A) &\cong GL'(R) &= E(R) \\
K_2(A) &= H_2(E(A),\mathbb{Z}) &\cong H_2(E(R)),\mathbb{Z}) &= K_2(R) \\
St(A) &= Sch(A) &\cong Sch(R) &= St(R) .
\end{aligned}$$

Now we investigate what is occuring before passage to limits. First we notice that the restriction of the isomorphism α to $E_p(A)$ is an isomorphism onto $E_{np}(R)$. This relies on

$$E_{ij}(\Lambda) = \prod_{k,l=1}^{n} E_{(ikjl)}(\lambda_{kl}) ,$$

where $i \neq j$, Λ is the $n \times n$ matrix (λ_{kl}) and $(ikjl) = ((i-1)n + k , (j-1)n + l)$, with $(i-1)n + k \neq (j-1)n + l$, i.e. eith $i \neq j$ or $k \neq j$ and $i = j$.

$$E_{(ikjl)}(\lambda) = \begin{cases} \alpha(E_{ij}(\lambda e_{kl})) & \text{if } i \neq j \\ \alpha([E_{i,i+1}(\lambda e_{k1}), E_{i+1,i}(e_{1l})]) & \text{if } i = j \text{ and } k \neq l. \end{cases}$$

THEOREM 2. For any integer $p \geq 3$, we have an epimorphism

$$\tilde{\alpha} : St_p(A) \to St_{np}(R)$$

defined by the formula

$$\tilde{\alpha}(x_{ij}(\Lambda)) = \prod_{k,l=1}^{n} x_{(ikjl)}(\lambda_{kl}),$$

where $i \neq j$, $\Lambda = (\lambda_{kl})$ and $(ikjl) = ((i-1)n+k, (j-1)n+l)$.

PROOF: It is easy to check that the family of elements $\tilde{\alpha}(x_{ij}(\Lambda))$ satisfy Steinberg relations (i), (ii) and (iii). Therefore $\tilde{\alpha}$ can be extended into a group homomorphism. To show that $\tilde{\alpha}$ is onto, let us take an arbitrary generator $x_{(ikjl)}(\lambda)$ in $St_{np}(R)$ and we see that

$$x_{(ikjl)}(\lambda) = \begin{cases} \tilde{\alpha}(x_{ij}(\lambda e_{kl})) & \text{if } i \neq j \\ \tilde{\alpha}([x_{i,i+1}(\lambda e_{k1}), x_{i+1,i}(e_{1l})]) & \text{if } i = j,\ k \neq l. \end{cases}$$

It is easy to show for a semi-simple ring A that one has $K_2(A) = K_2(D_1) \times K_2(D_2) \times \cdots \times K_2(D_2)$, where D_i's are the division rings associated with A by the structure theorem. What can be said in the "finite" case?

K. Dennis informs me that he has proven a stability theorem for rings satisfying a "stable range condition." This means that $\mathrm{Ker}(\varphi_n) \cong K_2(R)$ for

n sufficiently large.

Therefore it follows that the homomorphism $\tilde{\alpha}$ of Theorem 2 is actually an isomorphism if p is sufficiently large. In particular we have:

PROPOSITION 3. For any semi-simple ring A, the following exact sequence

$$1 \to L_p(A) \to St_p(A) \xrightarrow{\varphi_p} E_p(A) \to 1 \quad (p \geq 5)$$

satisfies

$$L_p(A) \cong \prod_i K_2(D_i) \cong \prod_i L_p(D_i),$$

where D_i are the division rings associated with A by $A \cong M_{n_i}(D_i)$.

Furthermore we have

$$St_p(A) \cong \prod_i St_{n_i p}(D_i).$$

PROOF: The situation is clear for a simple ring $A = M_n(D)$ because for a division ring D the extension $St_n(D) \to E_n(D)$ is central for $n \geq 3$ and we get $L_n(D) \cong K_2(D)$. For the general case let us remark that the functors E_n, St_n and L_n preserve the product.

REMARK. Using the "stability" result of Dennis for a field F it is possible to give a simple proof of Matsumoto's theorem by dealing with 3×3 matrices only.

REFERENCES

[1] H. Bass, K-theory and stable algebra, Publ. Math. I.H.E.S. 22 (1964).

[2] K. Dennis and M. Stein, K_2 of discrete valuation rings, (to appear).

[3] M. A. Kervaire, Multiplicateurs de Schur et K-théorie, Essays on Topology and related Topics, dedicated to G. de Rham, Springer 1970.

[4] J. Milnor, Notes on algebraic K-theory, mimeographed, M.I.T. (1969).

[5] I. Schur, Uber die Darstellung der endlichen Gruppen durch gebrochene lineare Substitutionen, J. Crelle, 127 (1904).

[6] R. G. Swan, Algebraic K-theory, Lecture Notes in Mathematics 76, Springer 1968.

CLASS GROUPS OF ORDERS AND A MAYER-VIETORIS SEQUENCE

I. Reiner* and S. Ullom*

University of Illinois

INTRODUCTION. A classical problem of algebraic number theory is to describe the ideal class group of the ring of integers R of an algebraic number field K finite over $\mathbb{Q}$. Here, the ideal classes are R-isomorphism classes of fractional R-ideals in K. To generalize, replace K by a semisimple finite dimensional $\mathbb{Q}$-algebra A, and replace R by a $\mathbb{Z}$-order Λ in A. Fractional ideals are replaced by Λ-lattices X in A, that is, finitely generated left Λ-modules X with $\mathbb{Q}X = A$. We restrict our attention to locally free (rank 1) Λ-lattices X, that is, those for which $X_p \cong \Lambda_p$ for each rational prime p, the subscript denoting localization. In the case where $\Lambda = \mathbb{Z}G$ with G a finite group, the locally free (rank 1) Λ-lattices are precisely the rank 1 projectives [19].

To define "addition" of lattices, recall Steinitz's classical result that for I, J fractional R-ideals, one has $I \dotplus J \cong R \dotplus IJ$. Swan [19] generalized this by proving that given any locally free Λ-lattices X, X' in A, there is another such lattice X'' for which $X \dotplus X' \cong \Lambda \dotplus X''$. The class group $\mathcal{C}(\Lambda)$ is then defined as the abelian group generated by all symbols $[Y]$, Y a locally free Λ-lattice, with $[X] = [Y]$ if and only if $X \dotplus \Lambda \cong Y \dotplus \Lambda$, and with addition defined via $[X] + [X'] = [X'']$ where $X \dotplus X' \cong \Lambda \dotplus X''$. By the Jordan-Zassenhaus theorem, $\mathcal{C}(\Lambda)$ is finite. One can also define $\mathcal{C}(\Lambda)$ as $\ker \rho$, where ρ is the rank homomorphism on the Grothendieck group of the category of all finite rank locally free Λ-lattices.

*This research was partially supported by a contract with the National Science Foundation.

There is a natural surjection $\mathcal{C}(\Lambda) \to \mathcal{C}(\Lambda')$, where Λ' is a maximal $\mathbb{Z}$-order in A containing Λ. Denote by $\mathcal{D}(\Lambda)$ the kernel of this surjection. We show here that $\mathcal{D}(\Lambda)$ is a p-group whenever G is a p-group [16], and also that $\mathcal{D}(\Lambda)$ is very large for any large abelian group G of composite order [17]. The proof of this latter result makes use of a Mayer-Vietoris sequence. This sequence also enables one to calculate $\mathcal{C}(\mathbb{Z}G)$ for G metacyclic of order pq [10].

Fröhlich [7] investigated a type of Picard group Picent Λ, the multiplicative group consisting of isomorphism classes of invertible Λ-Λ-bimodules in A. For commutative Λ, one has $\mathcal{C}(\Lambda) \cong$ Picent Λ. In the section on Class Groups and Picard Groups of Orders we give the relation between $\mathcal{C}(\Lambda)$ and Picent Λ in the general case [8].

Many of the results of this article hold equally well for R-orders. General references are [15, 18, 20]. Some of the material presented here appears in modified form in [8, 10, 16, 17].

EXPLICIT FORMULAS FOR $\mathcal{C}(\Lambda)$ AND $\mathcal{D}(\Lambda)$. Keeping the notation of the preceding section, denote by C the maximal order in the center of A. We may write

$$A = \sum_{j=1}^{t}{}^{\bullet} A_j, \quad A_j = \text{simple algebra},$$

$$\text{center of } A = \sum_{j=1}^{t}{}^{\bullet} K_j, \quad K_j = \text{center of } A_j,$$

$$C = \sum_{j=1}^{t}{}^{\bullet} R_j, \quad R_j = \text{alg. int. } \{K_j\}.$$

Fix a nonzero integer f such that $f\Lambda' \subset \Lambda$, and set

$$\Lambda_f = \bigcap_{p \mid f} \Lambda_p = \{\lambda/r : \lambda \in \Lambda, r \in Z, (r,f) = 1\}.$$

Denote by $u(S)$ the group of invertible elements of a ring S. An element x of S is said to be prime to f if $x \in u(S_f)$.

Let $I(R_j,f)$ be the group of all fractional R_j-ideals in K_j prime to f, and put

$$I(C,f) = \prod_{j=1}^{t} I(R_j,f).$$

Let $N_j : A_j \to K_j$, $j = 1, \ldots, t$, be the reduced norm map. For example, if $A_j = M_n(K_j)$ (= matrix algebra over K_j), then N_j is the determinant map to K_j. Let $N : A \to \mathbb{Q}C$ be defined by $N = \sum^{\bullet} N_j$, and let

$$I(\Lambda) = \{\prod_{j=1}^{t} R_j N_j(x_j) : \sum x_j \in u(A_f)\}$$

$$= \{C \cdot N(x) : x \in u(\Lambda_f)\}.$$

A simple algebra B is a totally definite quaternion algebra if its center F is a totally real algebraic number field, and the completion of B at every infinite prime of F is the skewfield of real quaternions. We say that the $\mathbb{Q}$-algebra A satisfies the Eichler condition if no simple component of A is a totally definite quaternion algebra.

THEOREM (2.1). (Jacobinski [12].) _Let Λ be a $\mathbb{Z}$-order in a semi-simple $\mathbb{Q}$-algebra A satisfying the Eichler condition. The reduced norm map (on ideals) induces an isomorphism $\mathcal{C}(\Lambda) \cong I(C,f)/I(\Lambda)$. If Γ is a $\mathbb{Z}$-order in A containing Λ, the "change of rings" map $\mathcal{C}(\Lambda) \to \mathcal{C}(\Gamma)$ corresponds to the natural surjection:_

$$I(C,f)/I(\Lambda) \to I(C,f)/I(\Gamma).$$

REMARK. Whether or not A satisfies the Eichler condition, the matrix algebra $M_2(A)$ always does. The $\mathbb{Z}$-order $M_2(\Lambda)$ has the same class group as Λ, (see [12]) and so one may use the above theorem to calculate $C(\Lambda)$. Since $M_2(A)$ has the same center as A, one obtains [12] an isomorphism $C(\Lambda) \cong I(C,f)/J(\Lambda)$, where now

$$J(\Lambda) = \{\prod_{j=1}^{t} R_j N_j^*(x_j) : \sum x_j \in GL_2(\Lambda_f)\}.$$

Here, $N_j^* : M_2(A_j) \to K_j$, $j = 1, \ldots, t$, is the reduced norm map.

The preceding results show that there is a surjection $\mathcal{C}(\Lambda) \to \mathcal{C}(\Lambda')$, where Λ' is a maximal $\mathbb{Z}$-order in A containing Λ. Since $\mathcal{C}(\Lambda')$ is easily calculated, it remains to determine the kernel $D(\Lambda)$ of this surjection.

THEOREM (2.2). _Suppose that the following four conditions hold for some rational prime_ p :

(i) $p^m\Lambda' \subset \Lambda \subset \Lambda'$ _for some_ m.

(ii) _For each_ j, $1 \leq j \leq t$, _there is a unique prime ideal_ P_j _of_ R_j _above_ p.

(iii) _For each_ $x \in \Lambda'$ _prime to_ p, _there exists_ $w \in \Lambda$ _prime to_ p _and_ $u \in u(C)$, _such that_ $N(x)N(w) \equiv u (\mathrm{mod}\ C \cap \mathrm{rad}\ C_p)$.

(iv) A _satisfies the Eichler condition._

Then $D(\Lambda)$ _is a p-group._

PROOF: By (ii), the Jacobson radical of C_p is given by

$$\mathrm{rad}\ C_p = \dot{\sum} \mathrm{rad}(R_j)_{P_j} = \dot{\sum} P_j(R_j)_{P_j},$$

and thus

$$P = C \cap \mathrm{rad}\ C_p = \dot{\sum} P_j.$$

From (2.1) we have $D(\Lambda) \cong I(\Lambda')/I(\Lambda)$. Given an element $z \in I(\Lambda')$, we will show that $z^{p^b} \in I(\Lambda)$ for some b. Write $z = C \cdot N(x)$, $x \in u(\Lambda_p')$; it suffices to consider the case where $x \in \Lambda'$ is prime to p. We may further assume that $N(x) \equiv u \pmod{P}$ for some $u \in u(C)$, since we need only replace x by xw, with w chosen as in (iii); this replacement does not affect $z \bmod I(\Lambda)$.

Since A satsifies the Eichler condition, it follows from [3] that $x \equiv v \pmod{P\Lambda'}$ for some $v \in u(\Lambda')$. Thus $xv^{-1} \in 1 + P\Lambda'$, and $z = C \cdot N(x) = C \cdot N(xv^{-1})$. Hence by (i), $y = (xv^{-1})^{p^b} \in \Lambda$ for large b. Thus $z^{p^b} = C \cdot N(y)$, and $y \in \Lambda_p \cap u(\Lambda_p') = u(\Lambda_p)$; this shows that $z^{p^b} \in I(\Lambda)$, and completes the proof.

The next theorem generalizes a result of Fröhlich [5] on abelian p-groups.

THEOREM (2.3). <u>If</u> G <u>is an arbitrary finite</u> p-<u>group, then</u> $D(\mathbb{Z}G)$ <u>is a</u> p-<u>group</u>.

PROOF: Let $\Lambda = \mathbb{Z}G$, and let us verify that (i)-(iv) of (2.2) are valid. Since $|G| \cdot \Lambda' \subset \Lambda$, surely (i) holds. Suppose now that p is odd; then (see [4]) each K_j is of the form $\mathbb{Q}(\omega)$, for some p^a-th root of unity ω. Thus (ii) and (iv) automatically hold true, and it remains to check (iii). By [4], there is exactly one K_j equal to $\mathbb{Q}$, say K_1, and each other K_j is a cyclotomic extension of $\mathbb{Q}$. Let $x \in \Lambda'$ be prime to p, and write $x = \sum x_i$, $x_i \in A_i$. We may choose $r \in Z \subset \Lambda$ prime to p such that

$$N_1(x_1)N_1(r_1) = N_1(x_1)r \equiv 1 \pmod{P_1}.$$

For $j > 1$, we have $K_j = \mathbb{Q}(\omega)$, $\omega = p^a$-th root of 1, $a \geq 1$, and the cyclotomic units $\{(1 - \omega^c)/(1 - \omega) : c \text{ prime to } p\}$ occupy all of the nonzero residue classes in R_j/P_j. But then for each j, $1 \leq j \leq t$, $N_j(x_j)N_j(r_j)$ is congruent (mod P_j) to a unit of R_j. This establishes (iii) of (2.2), and completes the proof for odd p.

For $p = 2$, $\mathbb{Q}G$ may fail to satisfy the Eichler condition, so we apply (2.2) with $M_2(\Lambda)$, $M_2(\Lambda')$ in place of Λ, Λ'. Again conditions (i) and (iv) hold true. Further, (ii) is valid since each K_j is a subfield of some $\mathbb{Q}(\omega)$, $\omega = 2^a$-th root of 1 (see [4]). Finally, (iii) holds because the residue class of 1 in the only unit in C/P.

REMARKS.

1. When G is not a p-group, it may happen that $|D(\mathbb{Z} G)|$ is divisible by primes not dividing $|G|$. This occurs for G cyclic of order $2p^n$, for example (see [21]).

2. For R a ring of algebraic integers, and G a p-group, it is not necessarily true that $|D(RG)|$ is also a p-group [21].

3. For p odd, we can prove (2.3) without using Eichler's results [3], by relying on the extra information that each A_j is a full matrix algebra over K_j (see [4]).

A MAYER-VIETORIS SEQUENCE FOR CLASS GROUPS. We present here a Mayer-Vietoris sequence useful in computing $D(\Lambda)$, with Λ a $\mathbb{Z}$-order (see [17] for proofs). Let Λ be a fiber product of rings with unit:

(3.1)
$$\begin{array}{ccc} \Lambda & \longrightarrow & \Lambda_1 \\ \downarrow & & \downarrow \varphi_1 \\ \Lambda_2 & \xrightarrow{\varphi_2} & \bar{\Lambda} \end{array}$$

that is,

$$\Lambda = \{(\lambda_1, \lambda_2) : \lambda_i \in \Lambda_i,\ \varphi_1\lambda_1 = \varphi_2\lambda_2\}.$$

If either φ_1 or φ_2 is surjective, Milnor [14] (see also Bass [1]) proved the exactness of the sequence:

$$(3.2)\quad K_1(\Lambda) \to K_1(\Lambda_1) \dotplus K_1(\Lambda_2) \to K_1(\bar{\Lambda}) \to K_0(\Lambda) \to K_0(\Lambda_1) \dotplus K_0(\Lambda_2) \to K_0(\bar{\Lambda}).$$

We give a more useful version, under the following hypotheses: $A = \mathbb{Q}\Lambda$ satisfies the Eichler condition, $\bar{\Lambda}$ is a finite ring, φ_1 or φ_2 is surjective, Λ_i are $\mathbb{Z}$-orders, $A_i = \mathbb{Q}\Lambda_i$ is semisimple, $i = 1, 2,$. Let $u^*(\Lambda_i)$ denote the image of $u(\Lambda_i)$ in $u(\bar{\Lambda})$. Then the sequence

$$(3.3)\quad 1 \to u^*(\Lambda_1) \cdot u^*(\Lambda_2) \to u(\bar{\Lambda}) \xrightarrow{\partial} D(\Lambda) \xrightarrow{f} D(\Lambda_1) \dotplus D(\Lambda_2) \to 0$$

is exact. The map f is obtained from "change of rings" homomorphisms, while for each $u \in u(\bar{\Lambda})$, $\partial u[\Lambda u]$, where

$$\Lambda u = \{(\lambda_1, \lambda_2) : \lambda_i \in \Lambda_i \;,\; (\varphi_1 \lambda_1)u = \varphi_2 \lambda_2\}.$$

Note that Λu is a locally free Λ-lattice such that $\Lambda' \otimes_\Lambda \Lambda u \cong \Lambda'$.

REMARK. Whether or not A satisfies the Eichler condition, the sequence

$$(3.4)\quad 1 \to GL_2^*(\Lambda_1) \cdot GL_2^*(\Lambda_2) \to GL_2(\bar{\Lambda}) \to D(\Lambda) \to D(\Lambda_1) \dotplus D(\Lambda_2) \to 0$$

is exact, where $GL_2^*(\Lambda_i)$ is the image of $GL_2(\Lambda_i)$ in $GL_2(\bar{\Lambda})$.

As an application of (3.3), we prove

THEOREM (3.5). _As_ G _ranges over any collection of abelian groups of composite order such that_ $|G| \to \infty$, _also_ $|D(\mathbb{Z}G)| \to \infty$.

Fröhlich [5] originally showed, for G abelian of exponent dividing 6, that $|D(\mathbb{Z}G)|$ is very large when $|G|$ is large. For the case where G is an abelian p-group, p odd, Fröhlich [6] recently obtained an exact formula for the order of $D(\mathbb{Z}G)^-$, the skew-symmetric part of $D(\mathbb{Z}G)$ relative to the involution defined by "complex conjugation." In particular, he showed that $|D(\mathbb{Z}G)^-| \to \infty$ as $|G| \to \infty$, provided that $|G| \neq p$.

Galovich [9] and Kervaire-Murthy [13] have gotten more explicit information about $D(\mathbb{Z}G)$, for G a cyclic p-group, by using deeper results from algebraic number theory.

We shall prove (3.5) for the case where G is an abelian p-group, p not 2; the more general assertion is established in [17]. Let G_0 be an abelian group of order p^{n+1}, $n \geq 1$, let $\langle x \rangle$ be a cyclic direct factor of G_0 of order p^m, $m \geq 1$, and write

$$G_0 = \langle x \rangle \times H, \quad G = \langle x^p \rangle \times H.$$

Let ω = primitive p^m-th root of 1, and let

$$R = \mathbb{Z}[\omega], \quad \overline{\mathbb{Z}} = \mathbb{Z}/p\mathbb{Z} \cong R/(1-\omega)R.$$

For $\Lambda = \mathbb{Z}G_0$, set $y = x^{p^{m-1}} \in G_0$, and define

$$I = (y^{p-1} + y^{p-2} + \ldots + y + 1)\Lambda, \quad J = (y-1)\Lambda,$$

a pair of ideals of Λ. Since $I \cap J = 0$, there is a fiber product

$$\begin{array}{ccc} \Lambda & \longrightarrow & \Lambda/I \\ \downarrow & & \downarrow \\ \Lambda/J & \longrightarrow & \Lambda/(I+J), \end{array}$$

or more explicitly,

(3.6)
$$\begin{array}{ccc} \mathbb{Z}G_0 & \longrightarrow & RH \\ \downarrow & & \downarrow \\ \mathbb{Z}G & \longrightarrow & \overline{\mathbb{Z}}G. \end{array}$$

Define an involution c on $\mathbb{Z}G_0$ by $g \to g^{-1}$, $g \in G_0$. Then $\mathbb{Z}G_0$, I, and J are c-invariant, so c acts on the rings of (3.6) and all maps are c-linear. On $R \subset RH$, c induces complex conjugation. For any $\langle c \rangle$-module M, let $M^+ = \{m \in M : cm = m\}$. Recall two basic classical results:

(3.7) (Kummer [2, Chap, 3, §1]). <u>The unit group is given by</u> $u(R) = u^+(R) \times \langle \omega \rangle$.

(3.8) (Higman [11]). <u>Let</u> W = <u>group of roots of unity of</u> R. <u>Then the torsion subgroup of</u> $u(RH)$ <u>is</u> $W \times H$.

PROPOSITION (3.9). <u>Given</u> $u \in u(RH)$. <u>There exist</u> $\omega^k \in W$ <u>and</u> $h \in H$ <u>such that</u> $c(u \cdot h\omega^k) = u \cdot h\omega^k$.

PROOF: Enlarge R, if necessary, so that its quotient field K is a splitting field for H. Then KH splits into a sum of copies of K. For $u \in u(RH)$, the projection $\pi_i(u)$ in the i-th simple component of KH is a unit in R. But $\pi_i(u/cu) = \pi_i(u)/c(\pi_i u)$, so by (3.7) it follows that $\pi_i(u/cu)$ is a root of unity. Since this holds for each i, we see that u/cu has finite order. Hence, using (3.8) and the fact that p is odd, we may write $u/cu = \pm\, \omega^{-2k} h^{-2}$ for some $h \in H$. Under the map $RH \xrightarrow{\epsilon} R \to R/(1-\omega)R$, with ϵ the augmentation map, we have $u/cu \to 1$, $\pm\omega^{-2k}h^{-2} \to \pm 1$. Thus the plus sign must occur, and then $c(u \cdot h\omega^k) = u \cdot h\omega^k$, as desired.

There are obvious modifications of (3.8) and (3.9) with RH replaced by $\mathbb{Z}G$.

Continuing with the proof of (3.5), it follows from (3.3) and (3.6) that the sequence

$$u(\mathbb{Z}G) \times u(RH) \xrightarrow{\varphi} u(\mathbb{Z}G) \to D(\mathbb{Z}G_0) \to D(RH) + D(\mathbb{Z}G) \to 0$$

is exact. Therefore $|D(\mathbb{Z}G_0)|$ is a multiple of $|D(\mathbb{Z}G)| \cdot [u(\overline{\mathbb{Z}}G) : \operatorname{im} \varphi]$. To complete the proof, we show that

(3.10) p^r divides $[u(\overline{\mathbb{Z}}G) : \operatorname{im} \varphi]$, where $r = \frac{1}{2}(p^n - 1) - n$.

Let N be the augmentation ideal of $\overline{\mathbb{Z}}G$, and let $U = 1 + N$. Then the unit group $u(\overline{\mathbb{Z}}G)$ is a direct product $u(\overline{\mathbb{Z}}) \times U$. Let $\tilde{G}$ be the image of G in $\overline{\mathbb{Z}}G$. Clearly $\tilde{G} \cap U^+ = 1$, and an easy calculation gives $[U : \tilde{G}U^+] = p^r$. It follows from (3.9) that both $\varphi\{u(RH)\}$ and $\varphi\{u(\mathbb{Z}G)\}$ lie in $u(\overline{\mathbb{Z}}) \times (\tilde{G}U^+)$. This establishes (3.10), and then (3.5) follows for abelian p-groups by induction on n.

OPEN QUESTION. For nonabelian p-groups G, does $|D(\mathbb{Z}G)| \to \infty$ as $|G| \to \infty$?

CLASS GROUPS AND PICARD GROUPS OF ORDERS. As in [18], let Picent Λ be the multiplicative group consisting of isomorphism classes of invertible Λ-bimodules X in A, with multiplication given by $\otimes_\Lambda$. Such an X is then a finitely generated two-sided Λ-ideal in A, and "invertibility" means that there exists another such Λ-ideal Y in A for which $XY = YX = A$. When Λ is commutative, we have Picent $\Lambda \cong \mathcal{C}(\Lambda)$, and Picent Λ is the usual Picard group (see [1, 18]). In the general case, let Pic' Λ be the subgroup of Picent Λ consisting of classes of locally free Λ-ideals X; here, X is <u>locally free</u> if for each rational prime p, $X_p \cong \Lambda_p$ as left Λ_p-modules (subscript denotes localization).

Let C_0 denote the center of Λ. We quote

THEOREM (4.1). (Fröhlich [7]). <u>There is an exact sequence</u>

$$1 = \text{Picent } C_0 \to \text{Picent } \Lambda \to \prod_p \text{Picent } \Lambda_{\hat{p}} \to 1,$$

<u>where the subscript $\hat{p}$ indicates completion. In the product,</u> p <u>ranges over the</u>

finite set of primes at which $\Lambda_p^{\wedge}$ *is not a separable algebra over its center. The sequence remains exact if Picent is replaced by Pic'.*

THEOREM (4.2). (Fröhlich [7]). *Let* $\Lambda = \mathbb{Z}G$, G = *p-group. Then* $D(C_0)$ *is also a p-group.*

We now define a homomorphism

$$\theta : \mathrm{Pic}'\ \Lambda \to \mathcal{C}(\Lambda)$$

by $\theta(X) = [X]$, where (X) is the bimodule class of X, and $[X]$ its class as left Λ-module. We shall summarize the results of [8] concerning the kernel and cokernel of θ. Assume for simplicity that A satisfies the Eichler condition, and let

$$N(\Lambda) = \{x \in u(A) : x\Lambda x^{-1} = \Lambda\}.$$

The following sequence is exact:

$$1 \to \frac{N(\Lambda)}{u(\Lambda)u(QC)} \xrightarrow{\omega} \mathrm{Pic}'\ \Lambda \xrightarrow{\theta} \mathcal{C}(\Lambda),$$

where ω maps the coset containing the element x of $N(\Lambda)$ onto the class of the Λ-bimodule Λx.

To describe the cokernel of θ, we use the notation introduced at the beginning of the section on Explicit Formulas for $\mathcal{C}(\Lambda)$ and $\mathfrak{D}(\Lambda)$, and further let $(A_i : K_i) = m_i^2$. Then

$$\text{cokernel } \theta \cong \frac{\prod_{j=1}^{t} I(R_j, f)}{\prod_j I(R_j, f)^{m_j} \cdot \{(C \cdot N(x))^f : x \in N(\Lambda_f)\}},$$

where $(C \cdot N(x))^f$ denotes the part of the ideal $C \cdot N(x)$ prime to f. This formula simplifies greatly when Λ is a maximal order.

REFERENCES

[1] H. Bass, Algebraic K-theory, Math. Lecture Note Series, Benjamin, New York, 1968.

[2] Z. I. Borevich and I. R. Shafarevich, Number theory, Academic Press, New York, 1966.

[3] M. Eichler, Allgemeine Kongruenzklasseneinteilungen der Ideale einfachen Algebren über algebraischen Zahlkörpern und ihre L-Reihen, J. Reine Angew. Math. 179 (1938), 227-251.

[4] W. Feit, Characters of finite groups, Math. Lecture Note Series, Benjamin, New York, 1967.

[5] A. Fröhlich, On the classgroups of integral group rings of finite Abelian groups, Mathematika 16 (1969), 143-152.

[6] __________, On the classgroups of integral group rings of finite Abelian groups, II, Mathematika (to appear).

[7] __________, Picard groups (to appear).

[8] A. Fröhlich, I. Reiner, and S. Ullom, Class groups and Picard groups of orders (to appear).

[9] S. Galovich, Nibs and Pics, Ph.D. Thesis, Brown University, Providence, Rhode Island, 1972.

[10] S. Galovich, I. Reiner, and S. Ullom, Class groups for integral representations of metacyclic groups, Mathematika (to appear).

[11] G. Higman, The units of group rings, Proc. London Math. Soc. (2) 46 (1940), 231-248.

[12] H. Jacobinski, Genera and decompositions of lattices over orders, Acta Math. 121 (1968), 1-29.

[13] M. A. Kervaire and M. P. Murthy, On the projective class group of cyclic groups of prime power order (to appear).

[14] J. Milnor, Introduction to algebraic K-theory, Annals of Math. Studies #72, Princeton Univ. Press, Princeton, N.J., 1971.

[15] I. Reiner, A survey of integral representation theory, Bull. Amer. Math. Soc. 76 (1970), 159-227).

[16] I. Reiner and S. Ullom, Class groups of integral group rings, Trans. Amer. Math. Soc. (to appear).

[17] __________, A Mayer-Vietoris sequence for class groups, J. of Algebra (to appear).

[18] K. W. Roggenkamp and V. H. Dyson, Lattices over orders. I, II. Lecture Notes No. 115, 142, Springer-Verlag, 1970.

[19] R. G. Swan, Induced representations and projective modules, Ann. of Math. (2) 71 (1960), 552-578.

[20] R. G. Swan and E. G. Evans, K-theory of finite groups and orders, Lecture Notes No. 149, Springer-Verlag, 1970.

[21] S. Ullom, A note on the classgroup of integral group rings of some cyclic groups, Mathematika 17 (1970), 79-81.

LIE PROPERTIES IN MODULAR GROUP ALGEBRAS

S. K. Sehgal

University of Alberta

INTRODUCTION. In the study of group rings one has frequently to invoke the Lie structure which is as usual defined by $[\gamma, \mu] = \gamma\mu - \mu\gamma$. Relevent to this talk are the following two examples. For $R = KG$, the group ring G over the field K, defined by induction the Lie central and Lie derived series by

$$R^{[1]} = R,\ R^{[2]} = [R,R],\ \ldots,\ R^{[n]} = [R^{[n-1]},R]$$

$$D^1R = R,\ D^2R = [R,R],\ \ldots,\ D^nR = [D^{n-1}R, D^{n-1}R]$$

where $[A,B]$ stands for the K-subspace generated by all $[a,b]$, $a \in A$, $b \in B$.

EXAMPLE 1. In 1940 Professor Zassenhaus [11] used

$$L_n(G) = \sum_{ip^j \geq n} (\Delta^{[i]}(G))^{\underline{p^j}}$$

where $\Delta(G)$ is the augmentation ideal of KG and char $(K) = p > 0$ to introduce a central series of the group G. Here $S^{\underline{p^j}}$ stands for the K-subspace generated by all s^{p^j} for $s \in S$. This central series is called the Brauer-Jennings-Zassenhaus M-series of G and will be defined more precisely later on.

EXAMPLE 2. In the study of the isomorphism problem for modular group algebras we [9] observed that

$$\mathfrak{z}(KG)/\mathfrak{z}(KG) \cap (KG)^{[2]} = K(\mathfrak{z}(G))$$

where G is a finite p-group and $\mathfrak{z}(A)$ stands for the centre of A. This had been known to Ward [10] also.

We decided to investigate in detail the Lie properties of KG and to look for applications. This is a report on joint works with Parmenter, Passi and Passman.

LIE SOLVABILITY AND LIE NILPOTENCE. We say R is Lie nilpotent if $R^{[n]} = 0$ for some n and Lie solvable if $D^n R = 0$ for some n. It is apparent that R is Lie nilpotent or Lie solvable if and only if it satisfies certain polynomial identities. Thus for example $D^2 R = 0$ if and only if R satisfies the identity

$$[[\alpha_1,\alpha_2], [\alpha_3,\alpha_4]] = 0$$

which is multilinear of degree 4. To state our main result we need the following:

DEFINITION. For $p > 0$ we say a group A is p-abelian if A' the commutator subgroup of A is a finite p-group. Moreover, for convenience, we say A is 0-abelian if and only if it is abelian.

THEOREM 1 [6]. Let KG be the group ring of G over the field K with char $K = p \geq 0$. Then

(i) KG is Lie nilpotent if and only if G is p-abelian and nilpotent.

(ii) For $p \neq 2$, KG is Lie solvable if and only if G is p-abelian.

(iii) For $p = 2$, KG is Lie solvable if and only if G has a 2-abelian subgroup of index at most 2.

APPLICATIONS TO GROUP OF UNITS. For $R = KG$ we define inductively series of 2-sided ideals $R^{(n)}$ and $D^{(n)}R$ by

$$R^{(1)} = R,\ R^{(2)} = [R^{(1)},R]R,\ \ldots,\ R^{(n)} = [R^{(n-1)},R]R$$

$$D^{(1)}R = R,\ D^{(2)}R = [D^{(1)}R,\ D^{(1)}R]R,\ \dots,\ D^{(n)}R = [D^{(n-1)}R,\ D^{(n-1)}R]R\ .$$

Obviously for abelian groups $R^{[2]} = 0 = R^{(2)} = D^2R = D^{(2)}R$. For nonabelian groups G consider the statements

(a) $(KG)^{[n]} = 0$ for some $n > 1$.

(b) G is nilpotent, char $K = p > 0$ and G is a finite p-group.

(c) $(KG)^{(m)} = 0$ for some $m > 1$.

a) $\Rightarrow$ (b) follows from Theorem 1. Induction on $|G'|$ gives easily (b) $\Rightarrow$ (c). Also, (c) $\Rightarrow$ a) is obvious. Hence we have proved the first part of the next proposition; the second part is easy.

PROPOSITION 2.

(a) $(KG)^{[n]} = 0$ <u>for some</u> $n \Leftrightarrow (KG)^{(m)} = 0$ <u>for some</u> m.

(b) $D^n(KG) = 0$ <u>for some</u> $n \Leftrightarrow D^{(m)}(KG) = 0$ <u>for some</u> m.

Let U denote the group of units of $R = KG$. Then for $\alpha, \beta \in U$

$$(\alpha,\beta) - 1 = \alpha^{-1}\beta^{-1}\alpha\beta - 1 = \alpha^{-1}\beta^{-1}(\alpha\beta - \beta\alpha)\ .$$

We conclude from this that

$$\gamma_n(U) \subseteq 1 + R^{(n)}$$

where $\gamma_n(U)$ is the nth term in the lower central series of U. Also,

$$D_n(U) \subseteq 1 + D^{(n)}R$$

where $D_n(U)$ is the nth term in the derived series of U. Hence we have

PROPOSITION 3.

(a) KG <u>Lie nilpotent</u> $\Rightarrow$ U <u>is nilpotent</u>.

(b) KG <u>Lie solvable</u> $\Rightarrow$ U <u>is solvable</u>.

REMARK. If G is finite, the converse implication in (a) also holds. This is a theorem of Bateman and Coleman ([1], [2]). Actually one can prove

PROPOSITION 4. Let G be a finitely generated group and $\operatorname{char} K = p > 0$. Then KG is Lie nilpotent $\Leftrightarrow$ U is nilpotent with U' a p-group of bounded exponent. (There also holds a similar result for Lie solvability of KG.)

Regarding the units in KG, the following is a problem of Herstein.

PROBLEM: Is it true that for $\gamma, \mu \in KG$, $\gamma\mu = 1 \Rightarrow \mu\gamma = 1$?

This has been answered in the affirmative for $\operatorname{char} K = 0$ and is an open question in case $\operatorname{char} K = p > 0$. We give a sufficient condition using the above theory. Suppose $\gamma, \mu \in KG$ and $\gamma\mu = 1$. Then observe that

$$e = 1 - \mu\gamma = \gamma\mu - \mu\gamma \in \Delta^{[2]}(G) \quad \text{and} \quad e^2 = e .$$

Further,

$$e = e^n \in \Delta^{(n)}(G) \quad \text{and} \quad e \in \Delta^{(w)} = \bigcap_n \Delta^{(n)}(G)$$

Also,

$$e \in \Delta^{[2]}(G) \subset (\Delta G')R, \quad e \in \bigcap_n ((\Delta G')^n R) .$$

It follows that

$$e \in (\Delta G')^w R .$$

We therefore investigate conditions for $\Delta^{(w)}(G)$ and $\Delta^w(G)$ to be nilpotent and have the following

THEOREM 5 [5]. Let $\operatorname{char} K = p > 0$. Then

(a) $\Delta^{(w)}(G)$ <u>is nilpotent if and only if</u> $D_{(w)}(G) = \{x \in G \mid x - 1 \in \Delta^{(n)}(G)$ <u>for all</u> $n\}$ <u>is a finite</u> p-<u>group</u>.

(b) $\Delta^{w}(G)$ <u>is nilpotent if and only if</u> $D_{w}(G) = \{x \in G \mid x - 1 \in \Delta^{n}(G)$ <u>for all</u> $n\}$ <u>is a finite</u> p-<u>group</u>.

COROLLARY 6 [5]. <u>Suppose</u> $\gamma, \mu \in KG$ <u>such that</u> $\gamma\mu = 1$ <u>and</u> char $K = p > 0$. <u>Then</u> $\mu\gamma = 1$ <u>if</u> $D_w(G')$ <u>is a finite</u> p-<u>group</u>. <u>In particular, if</u> $D_{(w)}(G)$ <u>is a finite</u> p-<u>group then</u> $\mu\gamma = 1$.

THE ISOMORPHISM PROBLEM AND THE BRAUER-JENNINGS-ZASSENHAUS SERIES. For a group G, let $M_i(G)$ be the ith term in its Brauer-Jennings-Zassenhaus series which is defined inductively by

$$M_1(G) = G\,,\ M_i(G) = (G, M_{i-1}(G))\, M_{(i|p)}(G)^p \quad \text{for } i \geq 2$$

where $(i|p)$ is the least integer $\geq i|p$ and $(G, M_{i-1}(G))$ denotes the subgroup generated by all commutators $(x,y) = x^{-1}y^{-1}xy$, $x \in G$, $y \in M_{i-1}(G)$. It is known ([3], [8], [11]) that if char $K = p > 0$ then

$$M_n(G) = \{g \in G \mid g - 1 \in \Delta^n(G)\}$$

$$= \prod_{ip^j \geq n} G_i^{p^j}$$

where G_i is the ith term in the lower central series

$$G = G_1 \geq G_2 \geq \cdots \geq G_i \geq \cdots \quad \text{of } G.$$

Write

$$L_n(G) = \sum_{ip^j \geq n} (\Delta^{[i]}(G))^{\underline{p^j}}\,.$$

Then it is easy to see that

LEMMA 7. $M_n(G)/M_{n+1}(G) \cong L_n(G)/\Delta^{n+1}(G)$.

Consider the natural monomorphism

$$(*) \qquad 0 \to M_n(G)/M_{n+1}(G) \to \Delta^n(G)/\Delta^{n+1}(G)$$

given by $m + M_{n+1}(G) \to m - 1 + \Delta^{n+1}(G)$ for $m \in M_n(G)$. Since $\Delta^n(G)/\Delta^{n+1}(G)$ is a vector space over K the field of p-elements, the embedding splits over K and we have

$$\Delta^n(G)/\Delta^{n+1}(G) = \Delta(G,M_n) + \Delta^{n+1}(G)/\Delta^{n+1}(G) + K_n(G)/\Delta^{n+1}(G)$$

where $K_n(G)$ is a subspace of KG. Moreover, since $\Delta^{n+1}(G) \subset K_n(G) \subset \Delta^n(G)$, we can conclude that $K_n(G)$ is an ideal. It turns out that $M_n(G)/M_{n+2}(G)$ is isomorphic to the group of units of $M_n(G) + K_{n+1}(G)/K_{n+1}(G)$. Careful checking gives the following

THEOREM 8 [7]. <u>Let</u> $\mathbb{Z}_p(G) \cong \mathbb{Z}_p(H)$. <u>Then</u>

(i) $M_i(G)/M_{i+1}(G) \cong M_i(H)/M_{i+1}(H)$

(ii) $M_i(G)/M_{i+2}(G) \cong M_i(H)/M_{i+2}(H)$

<u>for all</u> $i \geq 1$.

COROLLARY 9 [7]. <u>Suppose</u> $\mathbb{Z}_p(G) \cong \mathbb{Z}_p(H)$ <u>and that</u> $M_3(G) = 1$. <u>Then</u> $G \cong H$.

Analogous to (*), in the integral case i.e. in the group ring $\mathbb{Z}G$ there exists for $n = 2$ the sequence

$$(**) \qquad 0 \to \frac{\Delta(G,G') + \Delta^3(G)}{\Delta^3(G)} \to \frac{\Delta^2(G)}{\Delta^3(G)} \overset{\alpha}{\to} \frac{\Delta^2(G/G')}{\Delta^3(G/G')} \to 0 .$$

We know that there exists $\tau: \frac{\Delta^2(G/G')}{\Delta^3(G/G')} \to \frac{\Delta^2(G)}{\Delta^3(G)}$ and that $\alpha \circ \tau = 2I$. We [7] are able to exploit this fact to obtain some results on the integral isomorphism problem if division by 2 is uniquely defined in $\Delta^2(G/G')/\Delta^3(G/G')$. The sequence (**) is known to split in case G is finitely generated. We do not know if it splits in general.

DIMENSION THEORY. Parmenter [4] and Sandling [8] have computed in part the dimension subgroups

$$D_{n,R}(G) = \{g \in G \mid g - 1 \in \Delta_R^n(G)\}$$

and Lie dimension subgroups

$$D_{(n),R}(G) = \{g \in G \mid g - 1 \in \Delta_R^{(n)}(G)\}$$

of R(G) in terms of those of $\mathbb{Z}(G)$ and $\mathbb{Z}_r(G)$. We unify and complete their work.

Let $\sigma(R)$ be the set of those primes p such that $p^n R = p^{n+1} R$ for some n. Denote be e the smallest such n. Our results are:

THEOREM 10 [5].

(i) _If characteristic of_ $R = 0$, _then_

$$D_{n,R}(G) = \prod_{p \in \sigma(R)} \{\tau_p(G \bmod D_{n,\mathbb{Z}}(G)) \cap D_{n,\mathbb{Z}_{p^e}}(G)\}.$$

Here $\tau_p(G \bmod N)$ _denotes the_ p-_torsion subgroup of_ G mod N. _If_ $\sigma(R)$ _is empty then the right hand side is to be interpreted as_ $D_{n,\mathbb{Z}}(G)$.

(ii) _If characteristic of_ $R = r > 0$, _then_ $D_{n,R}(G) = D_{n,\mathbb{Z}_r}(G)$.

THEOREM 11 [5].

(i) <u>If characteristic of</u> $R = 0$, <u>then</u>

$$D_{(n),R}(G) = \prod_{p \in \sigma(R)} G' \cap \{\tau_p(G \bmod D_{(n),\mathbb{Z}}(G)) \cap D_{(n),\mathbb{Z}_{p^e}}(G)\}.$$

<u>If</u> $\sigma(R)$ <u>is empty the right hand side is to be interpreted as</u> $D_{(n),\mathbb{Z}}(G)$.

(ii) <u>If characteristic of</u> $R = r > 0$, <u>then</u> $D_{(n),R}(G) = D_{(n),\mathbb{Z}_r}(G)$.

Both these results follow from a general theorem regarding polynomial ideals in $R(G)$. For details we refer you to [5].

REFERENCES

[1] J. M. Bateman, On the solvability of unit groups of group algebras, Trans. Amer. Math. Soc. 157 (1971) 73-86.

[2] J. M. Bateman and D. B. Coleman, Group algebras with nilpotent unit groups, Proc. Amer. Math. Soc. 19 (1968), 448-449.

[3] S. A. Jennings, The structure of the group ring of a p-group over a modular field, Trans. Amer. Math. Soc. 50 (1941), 175-185.

[4] M. M. Parmenter, On a theorem of Bovdi, Canad. J. Math. 23 (1971), 929-932.

[5] M. M. Parmenter, I.B.S. Passi and S. K. Sehgal, Polynomial ideals in group rings, to appear in Canadian J. Math.

[6] I. B. S. Passi, D. S. Passman and S. K. Sehgal, Lie solvable group rings, to appear in Canadian J. Math.

[7] I. B. S. Passi and S. K. Sehgal, Isomorphism of modular group algebras, Math. Z., Vol. 129 (1972), 65-73.

[8] R. Sandling, The modular group ring of p-groups, Ph.D. Thesis, University of Chicago, 1969.

[9] S. K. Sehgal, On the isomorphism of group algebras, Math. Z. 95 (1967) 71-75.

[10] H. N. Ward, Some results on the group algebra of a group over a prime field, Mimeographed notes for the Seminar on finite groups and realted topics at Harvard University, 1960-61.

[11] H. J. Zassenhaus, Ein verfahren jeder endlichen p-Gruppe einen Lie-Ring mit der charakteristik p zuzuordnen, Abh. Mat. Sem. Hamb. 13 (1940), 200-207.

K-THEORY AND ALGEBRAIC CORRESPONDENCES

Richard G. Swan

University of Chicago

INTRODUCTION. In order to give a non-geometric proof of the Riemann Hypothesis for curves, Roquette [2] gave an elementary treatment of the theory of algebraic correspondences, i.e. one which does not make use of algebraic geometry. A simplified version of Roquette's proof was given by Eichler [1]. The proof involves a number of ad hoc constructions which make it appear quite complicated. I will show here how this difficulty can be avoided by using the methods of algebraic K-theory. The core of the proof remains the same but the preliminary constructions can be made more transparent in this way. Very little actual K-theory is used. The main point is that the exposition can be simplified by adopting the point of view of algebraic K-theory.

1. THE RING OF CORRESPONDENCES. Let k be an algebraically closed field and let K and L be function fields of dimension 1 over k. It is well known [1] that the ring $R = K \otimes_k L$ is a Dedekind ring. This is a special case of the following result.

LEMMA 1.1. *Let* $K_1, \ldots, K_n$ *be function fields of dimension* $d_1, \ldots, d_n$ *over* k. *Then* $K_1 \otimes_k K_2 \otimes_k \cdots \otimes_k K_n$ *is a noetherian ring of global dimension* $\sum d_i - \max(d_i)$.

PROOF: Each K_i is the quotient field of a finitely generated k-algebra R_i. Therefore $\otimes K_i$ is a localization of $\otimes R_i$ and so is noetherian. The regularity is obvious geometrically: Choose R_i to be the affine ring of a non-singular affine variety X_i. Then $\otimes R_i$ is the affine ring of $\prod X_i$.

This is non-singular, so $\otimes R_i$ is regular, and therefore, so is $R = \otimes K_i$ by localization. An algebraic version of this argument is easily given. By standard results on regular rings, gl. dim R = Krull dim R. Let $\mathfrak{m}$ be a maximal ideal of R. Since $K_i \hookrightarrow R/\mathfrak{m}$ for all i, we see that tr. deg $R/\mathfrak{m} \geq$ max d_i. Since $R_{\mathfrak{m}}$ is a geometric local ring, Krull dim $R_{\mathfrak{m}}$ = tr. deg R - tr. deg $R/\mathfrak{m} \leq \sum d_i - \max d_i$. Finally, if $d = \max d_i$, all K_i can be embedded in an algebraically closed field L of transcendence degree d over k. This gives a map $R \to L$. If $\mathfrak{m}$ is the kernel of this, tr. deg $R/\mathfrak{m} = d$ so Krull dim $R_{\mathfrak{m}} = \sum d_i - d$.

Returning now to the Dedekind ring $R = K \otimes_k L$, we define the group C(K,L) of classes of correspondences between K and L to be cl(R), the ideal class group of R. Let $\mathfrak{T}(R)$ be the category of finitely generated torsion modules over R. Then cl(R) is the quotient of $K_0(\mathfrak{T}(R))$ by the subgroup generated by all elements of the form $[R/(a)]$ where $a \in R$, $a \neq 0$. This follows from the fact that $K_0(\mathfrak{T}(R))$ is isomorphic to the ideal group of R or from the exact sequence of a localization

(1) $$K_1(F) \to K_0(\mathfrak{T}(R)) \to K_0(R) \to K_0(F)$$

where F is the quotient field of R.

LEMMA 1.2. <u>A finitely generated torsion module over</u> $R = K \otimes_k L$ <u>is the same as a</u> K,L-<u>bimodule which is finite dimensional over</u> K <u>and over</u> L.

PROOF: An R-module is clearly a K,L-bimodule. A finitely generated torsion module over R has a finite filtration with factors of the form $R/\mathfrak{m}$, $\mathfrak{m} \neq 0$. It is sufficient to look at the modules of this form. Now $R/\mathfrak{m}$ is a field containing isomorphic copies of K and L. If $R/\mathfrak{m}$ has transcendence degree ≥ 2 over k, the map $R \to R/\mathfrak{m}$ must be injective so $\mathfrak{m} = 0$. Therefore tr. deg $R/\mathfrak{m} = 1$ and $R/\mathfrak{m}$ is algebraic over K and L. But $R/\mathfrak{m}$ as an extension of K is generated by L which in turn is finitely generated over k. Therefore $R/\mathfrak{m}$ is finitely generated and algebraic and hence finite over K. The same applies to L. The converse is trivial.

Using this result we can define a bilinear map

$$(2) \qquad K_0(\mathfrak{J}(K_1 \otimes_k K_2)) \times K_0(\mathfrak{J}(K_2 \otimes_k K_3)) \to K_0(\mathfrak{J}(K_1 \otimes_k K_3))$$

where K_1, K_2, K_3 are function fields of dimension 1 over k. The map is given by sending $([V],[W])$ to $V \otimes_{K_2} W$. Note that $V \otimes_{K_2} W$ is a K_1, K_3 bimodule using the action of K_1 on V and K_3 on W. It follows immediately from Lemma 1.2 that $V \otimes_{K_2} W$ is a finitely generated torsion module over $K_1 \otimes_k K_3$.

LEMMA 1.3. <u>The map</u> (2) <u>induces a bilinear map</u>

$$(3) \qquad C(K_1,K_2) \times C(K_2,K_3) \to C(K_1,K_3)$$

PROOF: As noted above, $C(K_i,K_j)$ is a quotient of $K_0(\mathfrak{J}(K_i \otimes_k K_j))$. Therefore it is sufficient to show that if $V \in \mathfrak{J}(K_1 \otimes K_2)$ and $W = (K_2 \otimes K_3)/(a)$ where $a \neq 0$ then $V \otimes_{K_2} W$ goes to 0 in $\mathrm{cl}(K_1 \otimes K_3)$ or, equivalently (by (1)), in $K_0(K_1 \otimes K_3)$. Apply $V \otimes_{K_2}-$ to the sequence

$$0 \to K_2 \otimes K_3 \xrightarrow[a]{} K_2 \otimes K_3 \to W \to 0 .$$

Since $V \otimes_{K_2} (K_2 \otimes K_3) = V \otimes_k K_3$, this gives

$$0 \to V \otimes K_3 \to V \otimes K_3 \to V \otimes_{K_2} W \to 0 .$$

But V is finitely generated as a left K_1 module by Lemma 1.2 so $V \otimes K_3$ is finitely generated as a $K_1 \otimes K_3$ module. It follows that $V \otimes_{K_2} W = 0$ in $K_0(K_1 \otimes K_3)$. A similar argument applies if $V = (K_1 \otimes K_2)/(a)$.

We can now define a preadditive category $\mathcal{C}$ by taking the objects to be the function fields of dimension 1 over k, setting $\mathrm{Hom}(K,L) = C(K,L)$, and defining the composition to be given by (3). The associativity follows immediately from the corresponding property of tensor products. The unit $1_K \in C(K,K)$ is given by $[K]$ where K is regarded as a K,K bimodule in the obvious way. In particular, $C(K,K)$ is an associative ring with unit.

Since any K,L-bimodule can be regarded as an L,K-bimodule, there is a canonical isomorphism $C(K,L) \approx C(L,K)$, written $\xi \mapsto \xi'$. This gives an isomorphism between $\mathcal{C}$ and its dual. Clearly $\xi'' = \xi$.

REMARK. To get an additive category (with finite direct sums), the set of objects should be enlarged to the set of all k-algebras of the form $K_1 \times K_2 \times \cdots \times K_n$ where $K_i \in \operatorname{obj} \mathcal{C}$.

2. DIVISORS. We now show how a correspondence determines a map of divisor groups. The construction resembles that used to define the Brauer decomposition map in the K-theory of finite groups.

We consider only valuations of K and L which are trivial on k. These are all discrete rank 1 valuations. If $\mathfrak{p}$ is one, let $\mathfrak{O}_\mathfrak{p}$ denote its valuation ring and let $\pi_\mathfrak{p}$ be a prime element of $\mathfrak{O}_\mathfrak{p}$, i.e. $\operatorname{ord}_\mathfrak{p} \pi_\mathfrak{p} = 1$. For convenience we write KL for $K \otimes_k L$ and $\mathfrak{O}_\mathfrak{p}\mathfrak{O}_\mathfrak{Q}$ for $\mathfrak{O}_\mathfrak{p} \otimes \mathfrak{O}_\mathfrak{Q} \subset KL$. Let $\kappa = \kappa_\mathfrak{p} = \pi_\mathfrak{p} \otimes 1$ and $\lambda = \lambda_\mathfrak{Q} = 1 \otimes \pi_\mathfrak{Q}$ in $\mathfrak{O}_\mathfrak{p}\mathfrak{O}_\mathfrak{Q}$.

Let V be a finitely generated torsion module over KL. By Lemma 1.2, there is a finite set $S \subset V$ which generates V as a left K-module and as a right L-module. Let $M \subset V$ be the $\mathfrak{O}_\mathfrak{p}\mathfrak{O}_\mathfrak{Q}$ submodule of V generated by S. Then,

(4) $\qquad M \subset V$ is a finitely generated $\mathfrak{O}_\mathfrak{p}\mathfrak{O}_\mathfrak{Q}$ module and $KM = V = ML$

Equivalently we have $M_\kappa = M[\kappa^{-1}] = V = M_\lambda = M[\lambda^{-1}]$.

The canonical map $\mathfrak{O}_\mathfrak{p} \to \mathfrak{O}_\mathfrak{p}/(\pi_\mathfrak{p}) \to k$ gives us a map $\mathfrak{O}_\mathfrak{p}\mathfrak{O}_\mathfrak{Q} \to k\mathfrak{O}_\mathfrak{Q}\ (= k \otimes \mathfrak{O}_\mathfrak{Q}) = \mathfrak{O}_\mathfrak{Q}$ so $\mathfrak{O}_\mathfrak{Q}$ can be regarded as a quotient of $\mathfrak{O}_\mathfrak{p}\mathfrak{O}_\mathfrak{Q}$. In fact $\mathfrak{O}_\mathfrak{Q} = \mathfrak{O}_\mathfrak{p}\mathfrak{O}_\mathfrak{Q}/(\kappa)$.

Now $M \otimes_{\mathfrak{O}_p \mathfrak{O}_{\mathfrak{D}}} \mathfrak{O}_{\mathfrak{D}} = M/\varkappa M$ is a finitely generated $\mathfrak{O}_{\mathfrak{D}}$-module. Also, since $M_\lambda = V$ we have $(M/\varkappa M)_\lambda = V/\varkappa V = 0$ so $M/\varkappa M$ is a torsion module over $\mathfrak{O}_{\mathfrak{D}}$. We wish to define a divisor $V(\mathfrak{p})$ on L by setting $\mathrm{ord}_{\mathfrak{D}}(V(\mathfrak{p})) = \ell(M/\varkappa M)$, the length of $M/\varkappa M$ as an $\mathfrak{O}_{\mathfrak{D}}$-module. In order to do this we must show that this length is independent of M. While doing this we will also show that (4) can be replaced by a weaker assumption.

LEMMA 2.1. <u>Let</u> V <u>be a finitely generated torsion module over</u> KL <u>and let</u> $M \subset V$ <u>be a finitely generated</u> $\mathfrak{O}_p\mathfrak{O}_{\mathfrak{D}}$<u>-module with</u> $KML = V$. <u>Then</u> M <u>satisfies</u> (4) <u>and</u> $\mathrm{ord}_{\mathfrak{D}}(V(\mathfrak{p})) = \ell(M/\varkappa M)$ <u>depends only on</u> $\mathfrak{p}$, $\mathfrak{D}$, <u>and</u> V.

PROOF: If M and M' satisfy the hypothesis, we can find N satisfying (4) with $M, M' \subset N$. It will clearly suffice to compare M with N. Let $X = N/M$. Then $X_{\varkappa\lambda} = N_{\varkappa\lambda}/M_{\varkappa\lambda} = V/V = 0$ so $\varkappa^a\lambda^b X = 0$ for some a and b. Now $N_\varkappa = V$ and $\varkappa^a\lambda^b : V \approx V$. Since $X_\varkappa$ is a quotient of $N_\varkappa$ it follows that $\varkappa^a\lambda^b : X_\varkappa \to X_\varkappa$ is onto. Since this map is 0, we see that $X_\varkappa = 0$. Similarly, $X_\lambda = 0$ so $M_\varkappa = N_\varkappa = V$ and $M_\lambda = N_\lambda = V$. Thus M satisfies (4). We have also shown that $\varkappa^c X = \lambda^d X = 0$ for some c and d. By filtering X we reduce to the case $\varkappa X = \lambda X = 0$, i.e. X is an $\mathfrak{O}_p\mathfrak{O}_{\mathfrak{D}}/(\varkappa,\lambda) = k$ module. Filtering again we reduce to the case $X = k$. Now $0 \to M \to N \to k \to 0$ gives

$$\mathrm{Tor}_1^{\mathfrak{O}_p\mathfrak{O}_{\mathfrak{D}}}(N, \mathfrak{O}_{\mathfrak{D}}) \to \mathrm{Tor}_1^{\mathfrak{O}_p\mathfrak{O}_{\mathfrak{D}}}(k, \mathfrak{O}_{\mathfrak{D}}) \to M/\varkappa M \to N/\varkappa N \to k \to 0 .$$

The sequence $0 \to \mathfrak{O}_p\mathfrak{O}_{\mathfrak{D}} \xrightarrow{\varkappa} \mathfrak{O}_p\mathfrak{O}_{\mathfrak{D}} \to \mathfrak{O}_{\mathfrak{D}} \to 0$ shows that $\mathrm{Tor}_1(N, \mathfrak{O}_{\mathfrak{D}}) = \ker$ of $\varkappa$ on $N = 0$ since $N \subset V$, while $\mathrm{Tor}_1(k, \mathfrak{O}_{\mathfrak{D}}) = k$. Therefore, $0 \to k \to M/\varkappa M \to N/\varkappa N \to k \to 0$ and the result follows.

For any valuation $\mathfrak{D}$ of L/k, there is a unique extension $K\mathfrak{D}$ to the quotient field of KL such that $K\mathfrak{D}$ is trivial on K. Its valuation ring $\mathfrak{O}_{K\mathfrak{D}}$ is the localization of $K\mathfrak{O}_{\mathfrak{D}} = K \otimes \mathfrak{O}_{\mathfrak{D}}$ at the prime ideal (λ) and λ is a prime element of $\mathfrak{O}_{K\mathfrak{D}}$. It is easy to see that $\mathfrak{O}_{K\mathfrak{D}} \cap (KL) = K\mathfrak{O}_{\mathfrak{D}}$ and

$\mathfrak{G}_{K\mathfrak{Q}}\lambda \cap KL = K(\mathfrak{G}_{\mathfrak{Q}}\lambda)$. Similarly, we can define $\mathfrak{p}L$ for $\mathfrak{p}$ a valuation of K. Since $\mathfrak{p}L$ and $K\mathfrak{Q}$ have $\varkappa$ and λ as prime elements, we can normalize the representation of any principal ideal (a) of KL by assuming that $\text{ord}_{\mathfrak{p}L} a = 0 = \text{ord}_{K\mathfrak{Q}} a$.

LEMMA 2.2. *Let* $V = KL/(a)$ *where* $\text{ord}_{\mathfrak{p}L} a = 0$. *Let* $b \in L$ *be the image of* a *under* $\mathfrak{G}_{\mathfrak{p}}L \to kL = L$. *Then for every* $\mathfrak{Q}$ *in* L *we have* $\text{ord}_{\mathfrak{Q}}(V(\mathfrak{p})) = \text{ord}_{\mathfrak{Q}} b - \text{ord}_{K\mathfrak{Q}} a$.

PROOF: Without changing V or $\text{ord}_{\mathfrak{Q}} b - \text{ord}_{K\mathfrak{Q}} a$, we can assume that $\text{ord}_{K\mathfrak{Q}} a = 0$. Let $M = \mathfrak{G}_{\mathfrak{p}\mathfrak{Q}}/(a)$. We claim that Lemma 2.1 applies to this. Clearly $KML = V$. Now $\varkappa$ is a monomorphism on M since $\varkappa c = ad$ implies $c/a = d/\varkappa \in KL$ and $\text{ord}_{\mathfrak{p}L}(d/\varkappa) = \text{ord}_{\mathfrak{p}L}(c/a) \geq 0$ (since $\text{ord}_{\mathfrak{p}L} a = 0$), so $\varkappa c \in (a)$ implies $c \in (a)$. The same is true for λ. If Y is the kernel of $M \to V$ then $KYL = 0$ so if $y \in Y$ we have $\varkappa^r \lambda^s y = 0$ and therefore $y = 0$. This shows that $Y = 0$. Finally $M/\varkappa M = \mathfrak{G}_{\mathfrak{p}}\mathfrak{G}_{\mathfrak{Q}}/(a, \varkappa) = \mathfrak{G}_{\mathfrak{Q}}/(b)$ has length $\text{ord}_{\mathfrak{Q}} b$.

COROLLARY 2.3. *For any* $\mathfrak{p}$ *and* V, $\text{ord}_{\mathfrak{Q}} V(\mathfrak{p}) = 0$ *for almost all* $\mathfrak{Q}$.

PROOF: If $0 \to V' \xrightarrow{i} V \xrightarrow{j} V'' \to 0$, choose $M \subset V$ and let $M' = M \cap V'$, $M'' = j(M)$. If M is large enough, all of these satisfy (4). Also $0 \to M'/\varkappa M' \to M/\varkappa M \to M''/\varkappa M'' \to 0$ so

$$(5) \qquad \text{ord}_{\mathfrak{Q}} V(\mathfrak{p}) = \text{ord}_{\mathfrak{Q}} V'(\mathfrak{p}) + \text{ord}_{\mathfrak{Q}} V''(\mathfrak{p}).$$

Using this, we reduce to the case $V = KL/I$. Let $a \in I$, $a \neq 0$ and let $W = KL/(a)$. By (5), $\text{ord}_{\mathfrak{Q}} V(\mathfrak{p}) \leq \text{ord}_{\mathfrak{Q}} W(\mathfrak{p})$ and the result follows from Lemma 2.2.

Note that (5) shows that $V(\mathfrak{p})$ depends only on $[V] \in K_0(\mathfrak{J}(KL))$. Therefore we can define $v(\mathfrak{p})$ for any $v \in K_0(\mathfrak{J}(KL))$.

We can now give still another way to define $V(\mathfrak{p})$. Using (5), we reduce to the case $V = KL/P$ where P is prime. Therefore V is a field finite over K

and over L. Any $\mathfrak{p}$ on K determines a divisor on V in the usual way (as in algebraic number theory). The restriction of this divisor to L is $V(\mathfrak{p})$. This is very easy to see if V is separable over K. In this case, let R be the integral closure of $\mathfrak{O}_{\mathfrak{p}}$ in V. Let $\mathfrak{P}_1, \ldots, \mathfrak{P}_n$ be the maximal ideals of R. Then $\mathfrak{p}R = \mathfrak{P}_1^{e_1} \ldots \mathfrak{P}_n^{e_n}$ and this, considered as a divisor, is the extension of $\mathfrak{p}$ to V. Let $\mathfrak{Q}_i = \mathfrak{P}_i/L$. We must show that $V(\mathfrak{p}) = \prod \mathfrak{Q}_i^{e_i}$ but this is just the length of $R/\mathfrak{p}R$ so the result follows from Lemma 2.1. In the inseparable case, R may not be finitely generated over $\mathfrak{O}_{\mathfrak{p}}$ but the same argument may be applied by representing R as a direct limit of rings R_i finite over $\mathfrak{O}_{\mathfrak{p}}$. Lemma 2.1 applies to R_i for all sufficiently large i. This gives an alternative proof of Corollary 2.3. It also shows that $\deg V(\mathfrak{p}) = d'(V)$ (defined in §3).

3. THE TRACE. We now use methods similar to those in §2 to define the trace $\sigma: C(K,K) \to Z$. Let $\tilde{\mathfrak{P}}$ be the kernel of the map $KK = K \otimes K \to K$ by $x \otimes y \to xy$. Let $\mathfrak{J}'_K$ be the category of all finitely generated torsion modules V over $R = KK$ which satisfy the condition

$$R/\tilde{\mathfrak{P}} \otimes_R V = 0. \tag{6}$$

I will write K for $R/\tilde{\mathfrak{P}}$ since there is only one natural way to view K as a KK-bimodule. Similarly, if $\mathfrak{O} = \mathfrak{O}_{\mathfrak{p}}$ for some $\mathfrak{p}$, we regard $\mathfrak{O}$ as an $\mathfrak{O}\mathfrak{O}$-module via the map $\mathfrak{O}\mathfrak{O} \to \mathfrak{O}$ by $x \otimes y = xy$.

For each prime $\mathfrak{p}$ of K we will define a linear map $f_{\mathfrak{p}}: K_0(\mathfrak{J}'_K) \to \mathbb{Z}$. Let $\mathfrak{O} = \mathfrak{O}_{\mathfrak{p}}$ and let $V \in \mathfrak{J}'_K \subset \mathfrak{J}(KK)$. Choose $M \subset V$ satisfying (4). Since $M \otimes_{\mathfrak{O}\mathfrak{O}} \mathfrak{O}$ localizes to $V \otimes_{KK} K = 0$, it must be a torsion module over $\mathfrak{O}$. Define $f_{\mathfrak{p}}(V) = \ell(M \otimes_{\mathfrak{O}\mathfrak{O}} \mathfrak{O})$ to be its length. To show this is well defined, we reduce to the case $0 \to M \to N \to k \to 0$ as in the proof of Lemma 2.1, and, as in that proof, deduce the exact sequence

$$0 \to k \to M \otimes_{\mathfrak{O}\mathfrak{O}} \mathfrak{O} \to N \otimes_{\mathfrak{O}\mathfrak{O}} \mathfrak{O} \to k \to 0 .$$

This requires the following result.

LEMMA 3.1. (1) $\mathrm{hd}_{\mathfrak{O}\mathfrak{O}}\, \mathfrak{O} = 1$.

(2) *If* M *satisfies* (4), *then* $\mathrm{Tor}_1^{\mathfrak{O}\mathfrak{O}}(M, \mathfrak{O}) = 0$.

(3) $\mathrm{Tor}_1^{\mathfrak{O}\mathfrak{O}}(k, \mathfrak{O}) = k$.

PROOF: We first determine all prime ideals of $\mathfrak{O}\mathfrak{O}$. Let $\tilde{\mathfrak{m}} = (\varkappa, \lambda)$. This is maximal and $\mathfrak{O}\mathfrak{O}/\tilde{\mathfrak{m}} = k$. If $\mathfrak{P} \neq \tilde{\mathfrak{m}}$ then $\varkappa \notin \mathfrak{P}$ or $\lambda \notin \mathfrak{P}$. Say $\varkappa \notin \mathfrak{P}$. Then $(\mathfrak{O}\mathfrak{O})_{\mathfrak{P}}$ is the localization of $K \otimes_k \mathfrak{O}$ at $K \otimes p$, i.e. $(\mathfrak{O}\mathfrak{O})_{\mathfrak{P}} = \mathfrak{O}_{Kp}$ as in §2. This is a discrete valuation ring. It follows that any $\mathfrak{O}\mathfrak{O}$-module has $\mathrm{hd} \leq 1$ locally except possibly at $\tilde{\mathfrak{m}}$. Now, the completion of $(\mathfrak{O}\mathfrak{O})_{\tilde{\mathfrak{m}}}$ is clearly a power series ring in two variables $k[[x,y]]$ (where $x = \varkappa$, $y = \lambda$). The completion of $\mathfrak{O}$ at $\tilde{\mathfrak{m}}$ is therefore $k[[x,y]]/(x - y)$ which has $\mathrm{hd} = 1$. This proves (1).

For (2), let $0 \to P \to F \to \mathfrak{O} \to 0$ be a projective resolution of $\mathfrak{O}$ over $\mathfrak{O}\mathfrak{O}$. This gives a commutative diagram

$$\begin{array}{ccccccccc} 0 \to & \mathrm{Tor}_1^{\mathfrak{O}\mathfrak{O}}(\mathfrak{O}, M) & \to & P \otimes_{\mathfrak{O}\mathfrak{O}} M & \to & F \otimes_{\mathfrak{O}\mathfrak{O}} M & \to & \mathfrak{O} \otimes_{\mathfrak{O}\mathfrak{O}} M & \to 0 \\ & & & \downarrow p & & \downarrow f & & & \\ 0 \to & \mathrm{Tor}_1^{KK}(K,V) & \to & KPK \otimes_{KK} V & \to & KFK \otimes_{KK} V & \to & K \otimes_{KK} V & \to 0 \end{array}$$

where the bottom sequence is the localization of the top one with respect to $\varkappa\lambda$. Now $M \subset V$ so p and f are injective. It follows that $\mathrm{Tor}_1^{\mathfrak{O}\mathfrak{O}}(\mathfrak{O}, M) \subset \mathrm{Tor}_1^{KK}(K,V) = 0$ since $V \in \mathfrak{T}'_K$ so that locally one of K and V is always 0.

Finally, note that $k_{\mathfrak{P}} = 0$ for $\mathfrak{P} \neq \tilde{\mathfrak{m}}$. Therefore we can localize at $\tilde{\mathfrak{m}}$ and complete before calculating $\mathrm{Tor}(k, \mathfrak{O})$. Now $0 \to k[[x,y]] \xrightarrow{x-y} k[[x,y]] \to \hat{\mathfrak{O}} \to 0$ is a projective resolution and (3) follows immediately.

In the following, $K\mathfrak{p}$ and $\mathfrak{p}K$ have the same meaning as in §2.

LEMMA 3.2. <u>Let</u> $V = KK/(a) \in \mathfrak{T}'_K$ <u>and let</u> b <u>be the image of</u> a <u>under</u> $KK \to K$. <u>Then</u> $f_{\mathfrak{p}}(V) = \mathrm{ord}_{\mathfrak{p}}(b) - \mathrm{ord}_{\mathfrak{p}K}(a) - \mathrm{ord}_{K\mathfrak{p}}(a)$.

PROOF: Note that $b \neq 0$ since $V \in \mathfrak{T}'_K$. As in the proof of Lemma 2.2, we normalize a so that $\mathrm{ord}_{\mathfrak{p}K}(a) = 0 = \mathrm{ord}_{K\mathfrak{p}}(a)$ and conclude that in this case $M = \mathcal{O}\mathcal{O}/(a) \subset V$ and satisfies (4). Therefore $f_{\mathfrak{p}}(V)$ is the length of $M \otimes_{\mathcal{O}\mathcal{O}} \mathcal{O} = \mathcal{O}/(b)$, i.e. $f_{\mathfrak{p}}(V) = \mathrm{ord}_{\mathfrak{p}}(b)$.

COROLLARY 3.3. <u>For any</u> $V \in \mathfrak{T}'_K$, $f_{\mathfrak{p}}(V) = 0$ <u>for almost all</u> $\mathfrak{p}$.

PROOF: This follows in the same way as Corollary 2.3. If $0 \to V' \to V = V'' \to 0$, we can assume $0 \to M' \to M \to M'' \to 0$. Using Lemma 3.1 we conclude that

(7) $$f_{\mathfrak{p}}(V) \to f_{\mathfrak{p}}(V') + f_{\mathfrak{p}}(V'') .$$

For any KL-module V, define $d'(V) = \dim_K V$ and $d''(V) = \dim_L V$.

LEMMA 3.4. <u>If</u> $V = KL/(a)$ <u>with</u> $a \neq 0$, <u>then</u> $\sum_{\mathfrak{p} \text{ in } K} \mathrm{ord}_{\mathfrak{p}L}(a) = -d''(V)$ <u>and</u> $\sum_{\mathfrak{P} \text{ in } L} \mathrm{ord}_{K\mathfrak{P}}(a) = -d'(V)$.

PROOF: Let F be the quotient field of KL. The valuations of F which are trivial on K are the $K\mathfrak{P}$ and the valuations coming from primes P of KL. Since the $K\mathfrak{P}$ have degree 1 over K and F/K is a 1-dimensional function field we have

$$\sum_{\mathfrak{P} \text{ in } L} \mathrm{ord}_{K\mathfrak{P}}(a) + \sum_{P \text{ in } KL} |KL/P : K| \, \mathrm{ord}_P(a) = 0 .$$

The right hand sum is easily seen to be $d'(V)$ since KL is a Dedekind ring.

We now define the trace function σ by

$$\sigma(V) = d'(V) = d''(V) - \sum f_{\mathfrak{p}}(V)$$

for $V \in \mathfrak{T}'_K$, the sum being over all $\mathfrak{p}$ in K. It follows from Lemmas 3.2 and 3.4 that $\sigma(V) = 0$ if $V = KK/(a) \in \mathfrak{T}'_K$. By (7), σ is additive on short exact sequences. Therefore $\sigma : K_0(\mathfrak{T}'_K) \to \mathbb{Z}$ and is 0 on all $[KK/(a)]$. Now $K_0(\mathfrak{T}'_K)$ can be identified with the group of ideals of KK prime to $\tilde{\mathfrak{D}} = \ker[KK \to K]$. Under this identification, the subgroup A generated by all $[KK/(a)]$ for $KK/(a) \in \mathfrak{T}'_K$ becomes the group of principal ideals prime to $\tilde{\mathfrak{D}}$. Since KK is a Dedekind ring the quotient group is immediately seen to be $cl(KK)$, i.e. $K_0(\mathfrak{T}'_K)/A = C(K,K)$ so $\sigma : C(K,K) \to \mathbb{Z}$.

If $K \subset L$, we get a map $C(K,K) \to C(L,L)$ by sending V to $L \otimes_K V \otimes_K L$.

LEMMA 3.5. <u>If</u> $W = L \otimes_K V \otimes_K L$ <u>then</u> $\sigma(W) = |L:K|\sigma(V)$.

PROOF: Clearly $d'(W) = |L:K|d'(V)$ and similarly for d''. Let $\mathfrak{O} = \mathfrak{O}_{\mathfrak{p}}$ in K and let $\tilde{\mathfrak{O}}$ be the integral closure of $\mathfrak{O}$ in L. Then $\tilde{\mathfrak{O}} = \cap \tilde{\mathfrak{O}}_{\mathfrak{P}_i}$ over those $\mathfrak{P}_i$ extending $\mathfrak{p}$. Choose $M \subset V$ satisfying (4) and let $N = \tilde{\mathfrak{O}} \otimes_{\mathfrak{O}} M \otimes_{\mathfrak{O}} \tilde{\mathfrak{O}}$. Then $LNL = W$. To show that $N \subset W$ it will suffice to show it is torsion free over $\tilde{\mathfrak{O}}$ from the left and from the right (cf. proof of Lemma 2.2). Now M is torsion free over $\mathfrak{O}$ and so as an $\mathfrak{O}$-module is a direct limit of finitely generated free modules. For a free $\mathfrak{O}$-module F it is clear that $\tilde{\mathfrak{O}} \otimes_{\mathfrak{O}} F$ is torsion free over $\tilde{\mathfrak{O}}$. Similarly, since $\tilde{\mathfrak{O}}$ and M are torsion free over $\mathfrak{O}$ we see that $\tilde{\mathfrak{O}} \otimes_{\mathfrak{O}} M$ is torsion free over $\mathfrak{O}$. Repeating the argument shows that N is torsion free so $N \subset W$. For any $\mathfrak{P}|\mathfrak{p}$, Lemma 2.1 applies to $\tilde{\mathfrak{O}}_{\mathfrak{P}} N \tilde{\mathfrak{O}}_{\mathfrak{P}}$. Therefore,

$$\sum_{\mathfrak{P}|\mathfrak{p}} f_{\mathfrak{P}}(W) = \sum_{\mathfrak{P}|\mathfrak{p}} \ell(\tilde{\mathfrak{O}}_{\mathfrak{P}} N \tilde{\mathfrak{O}}_{\mathfrak{P}} \otimes_{\tilde{\mathfrak{O}}_{\mathfrak{P}}\tilde{\mathfrak{O}}_{\mathfrak{P}}} \tilde{\mathfrak{O}}_{\mathfrak{P}}) = \sum_{\mathfrak{P}|\mathfrak{p}} \ell(N \otimes_{\tilde{\mathfrak{O}}\tilde{\mathfrak{O}}} \tilde{\mathfrak{O}}_{\mathfrak{P}}) =$$

$$\sum_{\mathfrak{P}|\mathfrak{p}} \ell((N \otimes_{\mathfrak{O}\mathfrak{O}} \mathfrak{O}) \otimes_{\mathfrak{O}} \tilde{\mathfrak{O}}_{\mathfrak{P}}) = f_{\mathfrak{p}}(V) \sum_{\mathfrak{P}|\mathfrak{p}} \ell(\tilde{\mathfrak{O}}_{\mathfrak{P}} \otimes_{\mathfrak{O}} \mathfrak{O}/\mathfrak{p}) = |L:K| f_{\mathfrak{p}}(V) \quad \text{since}$$

$$\tilde{\mathfrak{O}}/\mathfrak{p}\tilde{\mathfrak{O}} = \prod_{\mathfrak{P}|\mathfrak{p}} \tilde{\mathfrak{O}}_{\mathfrak{P}}/\mathfrak{p}\tilde{\mathfrak{O}}_{\mathfrak{P}} \quad \text{and} \quad |\tilde{\mathfrak{O}}/\mathfrak{p}\tilde{\mathfrak{O}} : \mathfrak{O}/\mathfrak{p}| = |L:K|.$$

We must now compute $\sigma(K)$ where $K = KK/\tilde{\mathfrak{D}}$. Following [1] we will do this using differentials on K. Since KK is Dedekind, $\tilde{\mathfrak{D}}/\tilde{\mathfrak{D}}^2 \approx KK/\tilde{\mathfrak{D}} = K$. Define $d : K \to \tilde{\mathfrak{D}}/\tilde{\mathfrak{D}}^2$ by $dx = x \otimes 1 - 1 \otimes x$. This is a non-trivial differential. Since the universal module of differentials Ω_K is well known to be 1-dimensional over K we see that $d : \Omega_K \approx \tilde{\mathfrak{D}}/\tilde{\mathfrak{D}}^2$. Now if $\mathfrak{O} = \mathfrak{O}_{\mathfrak{p}}$, let Δ be the kernel of $\mathfrak{O}\mathfrak{O} \to \mathfrak{O}$. Let $I \subset \Delta$ be the ideal generated by $\kappa - \lambda = \pi \otimes 1 - 1 \otimes \pi$. Since $k[\pi]$ is dense in $\mathfrak{O}$, we see easily that $\Delta \subset I + (1 \otimes \pi^n)$ for any n. Now $\mathfrak{O}\mathfrak{O}/\Delta^2$ is local (mod nilpotent elements it is $\mathfrak{O}$). Therefore in $\mathfrak{O}\mathfrak{O}/\Delta^2 + I$ we have $\bigcap(1 \otimes \pi)^n = 0$ so in $\mathfrak{O}\mathfrak{O}$, $\Delta = \bigcap[\Delta^2 + I + (1 \otimes \pi)^n] = \Delta^2 + I$. This shows that $I \to \Delta/\Delta^2$ is onto so $\pi \otimes 1 - 1 \otimes \pi$ generates Δ/Δ^2 as an $\mathfrak{O}$-module. Since it localizes to $\tilde{\mathfrak{D}}/\tilde{\mathfrak{D}}^2 \approx K$, we see that $\Delta/\Delta^2 \approx \mathfrak{O}$. Also $\Delta/\Delta^2 \subset \tilde{\mathfrak{D}}/\tilde{\mathfrak{D}}^2 = \Omega_K$ and Δ/Δ^2 is just the set of all differentials ω with $\text{ord}_{\mathfrak{p}}\omega \geq 0$. To see this, observe that $d\pi$ generates $\tilde{\mathfrak{D}}/\tilde{\mathfrak{D}}^2$ since it generates Δ/Δ^2. Now $\text{ord}_{\mathfrak{p}} x d\pi = \text{ord}_{\mathfrak{p}} x$ for $x \in K$. But $\Delta/\Delta^2 = \{x d\pi \mid x \in \mathfrak{O}\}$.

LEMMA 3.6. <u>If</u> K <u>has genus</u> g <u>then</u> $\sigma(K) = 2g$.

PROOF: Choose $x \in K$ such that dx generates $\tilde{\mathfrak{D}}/\tilde{\mathfrak{D}}^2$. Let $\delta x = x \otimes 1 - 1 \otimes x$ in K so $dx = \delta x \bmod \tilde{\mathfrak{D}}^2$. The exact sequence

$$0 \to \tilde{\mathfrak{D}}/(\delta x) \to KK(\delta x) \to KK/\tilde{\mathfrak{D}} \to 0$$

shows that $\sigma(K) = -\sigma(\tilde{\mathfrak{D}}/(\delta x))$. Now $\tilde{\mathfrak{D}}/(\delta x) \in \mathfrak{T}'_K$ so the definition of $f_{\mathfrak{p}}$ can be applied. Choose $y, z \in \mathfrak{O}$ such that $(y \otimes z)\delta x \in \mathfrak{O}\mathfrak{O}$ and has $\text{ord}_{\mathfrak{p}K} = \text{ord}_{K\mathfrak{p}} = 0$. As in the proof of Lemma 2.2 we can choose $M = \Delta/((y \otimes z)\delta x)$. Now $M \otimes_{\mathfrak{O}\mathfrak{O}} \mathfrak{O} = \Delta/[\Delta^2 + ((y \otimes z)\delta x)]$ whose length is $\text{ord}_{\mathfrak{p}}(yz\, dx)$. Since $\text{ord}_{\mathfrak{p}K}((y \otimes z)\delta x) = 0$ we have $\text{ord}_{\mathfrak{p}} y = -\text{ord}_{\mathfrak{p}K}(\delta x)$. Similarly, $\text{ord}_{\mathfrak{p}} z = -\text{ord}_{K\mathfrak{p}}(\delta x)$. Using Lemma 3.4 we see that

$$\sum_{\mathfrak{p}} f_{\mathfrak{p}}(\tilde{\mathfrak{D}}/(\delta x)) = \sum_{\mathfrak{p}} \text{ord}_{\mathfrak{p}}(dx) + d'(KK/(\delta x)) + d''(KK/(\delta x)).$$

Therefore $\sigma(K) = -\sigma(\tilde{\mathfrak{D}}/(\delta x)) = 2 + \sum_{\mathfrak{p}} \operatorname{ord}_{\mathfrak{p}}(dx) = 2 + 2g - 2 = 2g$ since (dx) is a canonical divisor. Note here that $d'(KK/(\delta x)) = 1 + d'(\tilde{\mathfrak{D}}/(\delta x))$ and similarly for d''.

4. ROQUETTE'S PROOF. The object is, of course, to prove the following result in which $\xi \mapsto \xi'$ is the canonical involution considered in §1.

THEOREM 4.1. _If_ $\xi \in C(K,L)$ _and_ $\xi \neq 0$ _then_ $\sigma(\xi\xi') > 0$.

A preliminary reduction is made by considering the quotient field F of KL as a function field over K. The valuations of F trivial on K and L come from the primes of KL. Those which are non-trivial on L have the form $K\mathfrak{O}$. Therefore the divisor group of F/K can be identified as

$$(8) \qquad \mathfrak{D}_{F/K} = K_0(\pi(KL)) \times \mathfrak{D}_{L/k}$$

where $\mathfrak{D}_{L/K}$ is the divisor group of L/k. If $D_1 = (v_1, \mathfrak{A}_1) \sim D_2 = (v_2, \mathfrak{A}_2)$, then v_1 and v_2 clearly determine the same element of $C(K,L)$. We now extend the notation of §2. If $D \in \mathfrak{D}_{F/K}$ with $D = (v, \mathfrak{A})$ in (8), we define $D(\mathfrak{p}) = v(\mathfrak{p}) + \mathfrak{A}$ for any valuation $\mathfrak{p}$ of K/k. It follows immediately from Lemma 2.2 that $D \sim 0$ implies $D(\mathfrak{p}) \sim 0$ for all $\mathfrak{p}$. It is clear that $D \geq 0$ implies $D(\mathfrak{p}) \geq 0$. Also $\deg D(\mathfrak{p}) = \deg D$ by the remark at the end of §2. We will need one further observation. Suppose $K_1 \supset K$. Let F_1 be the quotient field of K_1L and define a map $\mathfrak{D}_{F/K} \to \mathfrak{D}_{F_1/K_1}$ by sending $D = ([V], \mathfrak{A})$ to $D_1 = ([K_1 \otimes_K V], \mathfrak{A})$ (using (8)). This agrees with the usual map, of course. Let $\mathfrak{p}$ be a valuation of K and let $\mathfrak{p}_1$ be an extension of $\mathfrak{p}$ to K_1. The fact we need is that

$$(9) \qquad D_1(\mathfrak{p}_1) = D(\mathfrak{p}).$$

This is obvious for the $\mathfrak{A}$ term. Choose $M \subset V$ satisfying (4) for $\mathfrak{O}_{\mathfrak{p}}\mathfrak{O}_{\mathfrak{Q}}$. Let $N = \mathfrak{O}_{\mathfrak{p}_1} \otimes_{\mathfrak{O}_{\mathfrak{p}}} M$. As in the proof of Lemma 3.5, this is torsion free and Lemma 2.1 applies. Therefore $\operatorname{ord}_{\mathfrak{Q}} V_1(\mathfrak{p}_1) = \ell(k \otimes_{\mathfrak{O}_{\mathfrak{p}_1}} N) = \ell(k \otimes_{\mathfrak{O}_{\mathfrak{p}}} M) = \operatorname{ord}_{\mathfrak{Q}} V(\mathfrak{p})$.

We can now perform the preliminary reduction following the procedure in [1]. Let η be a divisor of L such that

(10) $\qquad \eta \geq 0$, $\deg \eta = g_L$, and $\dim(\mathcal{K} - \eta) = 0$

where g_L is the genus of L and $\mathcal{K}$ is a canonical divisor of L. Fix $\mathfrak{p}$ in K. Given v representing ξ, let $\mathfrak{A} = v(\mathfrak{p})$ and set $D = (v, -\mathfrak{A} + \eta)$. Then $D(\mathfrak{p}) = \eta$ so $\deg D = \deg \eta = g_L = g_{F/K}$. By the Riemann-Roch theorem for F/K, $D \sim D'$ where $D' \geq 0$. Also $D'(\mathfrak{p}) \geq 0$ and $D'(\mathfrak{p}) \sim D(\mathfrak{p}) = \eta$. Now $\dim \eta = 1$ by Riemann-Roch and (10) so $D'(\mathfrak{p}) = \eta$. Write $D' = (u, \mathfrak{B})$. Since $D' \geq 0$, $u = [V]$ for some KL-module V. We can assume that $V = V_1 \times \cdots \times V_r$ where V_i has the form KL/P_i. If $d'(V_i) > 1$ we can extend K to $K' = V_i$. This causes $K' \otimes_K V_i$ to become reducible as a $K'L$-module. Repeating this process, we eventually obtain an extension K_1 of K such that $K_1 \otimes_K V$ has a composition series all of whose factors have $d' = 1$. By (9), this process does not change $V(\mathfrak{p}) = \eta'$ which satisfies $\eta' \leq \eta$. Also, $\sigma(\xi\xi')$ is just multiplied by $|K_1 : K|$.

LEMMA 4.2. __It is sufficient to prove Theorem 4.1 for the case where__ ξ __is represented by a__ KL-__module__ V __of the form__ $V = V_1 \times \cdots \times V_r$ __where each__ V_i __has the form__ $V_i = KL/P_i$ __and__ $d'(V_i) = 1$. __We can also assume that for a fixed__ $\mathfrak{p}_0$ __in__ K, __we have__ $V_i(\mathfrak{p}_0) = \mathfrak{Q}_i$ __where__ $\mathfrak{Q}_1, \ldots, \mathfrak{Q}_r$ __is a non-special system of prime divisors of__ L __with__ $r \leq g_L$.

The non-speciality of the $\mathfrak{Q}$'s means that $\sum \mathfrak{Q}_i \leq \eta$ where η satisfies (10). Note that V_i is a field of degree 1 over K. Therefore we can identify V_i with K, the K-action being the usual one and the L-action being given by an embedding $\varphi_i : L \to K$. The remarks at the end of §2 show that $V_i(\mathfrak{p})$ is simply the restriction of $\mathfrak{p}$ to L with respect to this embedding and, in particular, is prime.

It is easy to compute $\sigma(\xi\xi')$ if ξ has the form given in Lemma 4.2. If $\varphi : L \to K$ is an embedding (with $\varphi | k = \mathrm{id}$), let $V_\varphi = K$ with the usual K-action and with L acting through φ. Let V'_φ be V_φ as an LK-module. Let $\xi_\varphi \in C(K,L)$ be the class of V_φ. For $\mathfrak{p}$ in K we write $\varphi^{-1}\mathfrak{p}$ for the restriction of $\mathfrak{p}$ to L using the embedding φ.

LEMMA 4.3. (1) $\sigma(\xi_\varphi \xi'_\varphi) = 2g_L |K : \varphi L|$

(2) <u>If</u> $\varphi \neq \psi$ <u>and</u> $\varphi^{-1}\mathfrak{p} \neq \psi^{-1}\mathfrak{p}$, <u>then</u> $f_\mathfrak{p}(V_\varphi \otimes_L V'_\varphi) = 0$.

(3) <u>If</u> $\varphi \neq \psi$ <u>and</u> $\varphi^{-1}\mathfrak{p} = \psi^{-1}\mathfrak{p} = \mathfrak{O}$, <u>then</u> $f_\mathfrak{p}(V_\varphi \otimes_L V'_\varphi) =$

$\ell(\mathfrak{O}_\mathfrak{p} / (\varphi(x) - \psi(x), \text{ all } x \in \mathfrak{O}_\mathfrak{O})) = \mathrm{ord}_\mathfrak{p}(\varphi(\pi) - \psi(\pi))$

<u>where</u> π <u>is a prime element of</u> $\mathfrak{O}_\mathfrak{O}$.

PROOF: In (1) we have $V_\varphi \otimes_L V'_\varphi = K \otimes_L K = K \otimes_L L \otimes_L K$ and the result follows from Lemmas 3.5 and 3.6. In (2) and (3) we compute $f_\mathfrak{p}$ by choosing $M \subset V_\varphi \otimes_L V'_\otimes = K \otimes_L K$ to be the image of $\mathfrak{O}\mathfrak{O} = \mathfrak{O} \otimes \mathfrak{O} \subset K \otimes K$ where $\mathfrak{O} = \mathfrak{O}_\mathfrak{p}$. If $M = \mathfrak{O}\mathfrak{O}/I$ then $M \otimes_{\mathfrak{O}\mathfrak{O}} \mathfrak{O} = \mathfrak{O}/J$ where J is the image of I in $\mathfrak{O}$. In case (2), since $\varphi^{-1}\mathfrak{p} \neq \psi^{-1}\mathfrak{p}$ we can find $a \in L$ such that $\mathrm{ord}_{\varphi^{-1}\mathfrak{p}} a = 0$ and $\mathrm{ord}_{\psi^{-1}\mathfrak{p}} a > 0$. Then $\mathrm{ord}_\mathfrak{p} \varphi a = 0$ while $\mathrm{ord}_\mathfrak{p} \psi a > 0$. But $\varphi(a) \otimes 1 - 1 \otimes \psi(a) \in I$ so $\varphi(a) - \psi(a) \in J$. Thus $J = \mathfrak{O}$. In case (3), note that $M = \mathfrak{O} \otimes_{\mathfrak{O}_\mathfrak{O}} \mathfrak{O}$ since this is torsion free as in the proof of Lemma 3.5. Therefore I is generated by all $\varphi(x) \otimes 1 - 1 \otimes \psi(x)$ for $x \in \mathfrak{O}_\mathfrak{O}$. For the last equality we need only note that $k[\pi]$ is dense in $\mathfrak{O}_\mathfrak{O}$.

Everything is now very explicit and the method used by Roquette can be applied to complete the proof of Theorem 4.1. I have nothing new to add to this part of the proof. However, since this paper is essentially expository in character, I will repeat the details here for the reader's convenience.

Using Lemma 3.4, we see that $\sigma(\xi\xi') = 2g\sum |K : \varphi_i(L)| + 2\sum_{i<j} \sigma(\xi_i\xi_j')$

where ξ_i is the class of V_i and $g = g_L$. Let $W_{ij} = V_i \otimes_L V_j'$ and $W = \prod_{i<j} W_{ij}$. Then $d'(W) = \frac{1}{2}\sum_{i\neq j} |K : \varphi_j L| = \frac{1}{2}d(r-1)$ where $d = \sum |K : \varphi_i(L)|$.

The same value is obtained for $d''(W)$. Therefore,

(11) $$\sigma(\xi\xi') = 2dg + 2d(r-1) - 2\sum_{\mathfrak{p}} f_{\mathfrak{p}}(W).$$

To estimate the last sum we use the following result as in [1] and [2].

LEMMA 3.5. <u>Let</u> $\mathfrak{D}_1, \ldots, \mathfrak{D}_r$, $r \leq g$ <u>be a non-special system of divisors in</u> L. <u>Then we can find</u> $w_1, \ldots, w_r \in L$ <u>such that</u> $w_i \equiv \delta_{ij} \bmod \mathfrak{D}_j$, <u>and such that the denominators</u> $(w_i)_\infty$ <u>of the</u> (w_i) <u>satisfy</u> $0 \leq (w_i)_\infty \leq \mathcal{Y}$ <u>where</u> $\deg \mathcal{Y} \leq g + r - 2$. <u>We can also assume that</u> $\mathcal{Y}$ <u>is prime to any given divisor</u>.

The proof is given below.

We choose $\mathcal{Y}$ prime to all $\varphi_i^{-1}(\mathfrak{p})$ for all $\mathfrak{p}$ with $f_{\mathfrak{p}}(W) \neq 0$. Let $\Delta = \det|\varphi_i(w_j)|$. Then $\Delta \equiv 1 \bmod \mathfrak{p}_0$ so $\Delta \neq 0$. The donominator of (Δ) satisfies $(\Delta)_\infty \leq \sum \varphi_i(\mathcal{Y})$ where $\varphi_i(\mathcal{Y})$ is the extension of $\mathcal{Y}$ to K using φ_i. Now $\deg \varphi_i(\mathcal{Y}) = |K : \varphi_i L| \deg \mathcal{Y}$ so we have

(12) $$\deg(\Delta)_0 = \deg(\Delta)_\infty \leq d(g + r - 2).$$

Now let $\mathfrak{p}$ be a prime of K with $f_{\mathfrak{p}}(W) \neq 0$. Renumber the indices so that $\varphi_1^{-1}\mathfrak{p} = \cdots = \varphi_{s_1}^{-1}\mathfrak{p} \neq \varphi_{s_1+1}^{-1}\mathfrak{p} = \cdots = \varphi_{s_1+s_2}^{-1}\mathfrak{p} \neq \cdots$. We expand Δ using blocks of size $s_1, s_2, \ldots$, i.e. $\Delta = \sum \pm A_1A_2 \cdots$ where A_1 is an $s_1 \times s_1$ minor involving the first s_1 rows, A_2 is an $s_2 \times s_2$ minor involving rows

$s_1 + 1$ to $s_1 + s_2$, etc. To estimate A_1 we note that all $w_i \in \mathfrak{O}_{\varphi_1^{-1}\mathfrak{p}}$ since γ is prime to $\varphi_1^{-1}\mathfrak{p}$ by choice. Let π be prime for $\varphi_1^{-1}\mathfrak{p}$. Since $k[\pi]$ is dense in $\mathfrak{O}_{\varphi_1^{-1}\mathfrak{p}}$, we see that A_1 can be approximated at $\mathfrak{p}$ by a linear combination of determinants of the form

$$\begin{vmatrix} \varphi_1(\pi^{\nu_1}) & \varphi_1(\pi^{\nu_2}) & \cdots\cdots & \varphi_1(\pi^{\nu_{s_1}}) \\ \vdots & \vdots & & \vdots \\ \varphi_{s_1}(\pi^{\nu_1}) & \varphi_{s_1}(\pi^{\nu_2}) & \cdots\cdots & \varphi_{s_1}(\pi^{\nu_{s_1}}) \end{vmatrix}$$

Now any determinant of the form

$$\begin{vmatrix} x_1 & x_1 & \cdots & x_1 \\ \vdots & \vdots & & \vdots \\ x_m & x_m & \cdots & x_m \end{vmatrix}$$

is divisible by $\prod_{i<j} (x_i - x_j)$ in the ring $\mathbb{Z}[x_1, \ldots, x_m]$ since it vanishes if $x_i = x_j$. Therefore $\operatorname{ord}_{\mathfrak{p}} A_1 \geq \operatorname{ord}_{\mathfrak{p}} \prod_{i<j\leq s_1} (\varphi_i(\pi) - \varphi_j(\pi))$. Similar results hold for $A_2, A_3, \ldots$. Now $f_{\mathfrak{p}}(W_{ij}) = 0$ if $\varphi_i^{-1}\mathfrak{p} \neq \varphi_j^{-1}\mathfrak{p}$ while $f_{\mathfrak{p}}(W_{ij}) = \operatorname{ord}_{\mathfrak{p}}(\varphi_i(\pi) - \varphi_j(\pi))$ for $\varphi_i^{-1}\mathfrak{p} = \varphi_j^{-1}\mathfrak{p}$. Therefore $f_{\mathfrak{p}}(W) \leq \operatorname{ord}_{\mathfrak{p}}\Delta$ and so $\sum f_{\mathfrak{p}}(W) \leq \deg(\Delta)_0$. Substituting this in (11) and using (12) gives $\sigma(\xi\xi') \geq 2d > 0$. Of course if $g = 0$ or 1, the calculation with determinants can be avoided.

PROOF OF LEMMA 3.5 : Suppose γ is required to be prime to $\mathfrak{p}_1, \ldots, \mathfrak{p}_s$. We can clearly assume $r > 0$ and that $\mathfrak{O}_1, \ldots, \mathfrak{O}_r$ are included in $\mathfrak{p}_1, \ldots, \mathfrak{p}_s$. By the non-speciality, there are $\mathfrak{O}_{r+1}, \ldots, \mathfrak{O}_g$ such that $\dim(\mathcal{K} - \mathfrak{O}_1 - \ldots - \mathfrak{O}_g) = 0$. Since $\dim \mathcal{K} = g$, the dimension must drop by 1 as each successive $\mathfrak{O}$ is subtracted. Therefore $\mathfrak{A} = \mathcal{K} - \mathfrak{O}_{r+1} - \ldots - \mathfrak{O}_g$ has the following properties:

(a) $\deg \mathfrak{A} \leq g + r - 2$

(b) $\dim \mathfrak{A} = r$

(c) $\dim (\mathfrak{A} - \mathfrak{O}_1 - \ldots - \mathfrak{O}_r) = 0$

Any $\mathfrak{A}$ satisfying (b) has $\deg \mathfrak{A} \geq 0$ since $r > 0$. Choose $\mathfrak{A}$ satisfying (a), (b), (c) with $\deg \mathfrak{A}$ least. Then for each prime $\mathfrak{p}$, $\dim (\mathfrak{A} - \mathfrak{p}) < r$ otherwise $\mathfrak{A} - \mathfrak{p}$ would satisfy (a), (b), (c). Since k is infinite we can find $f \in \mathfrak{L}(\mathfrak{A})$ such that $f \notin \mathfrak{L}(\mathfrak{A} - \mathfrak{p}_i)$ for $i = 1, \ldots, s$. Let $\mathcal{Y} = \mathfrak{A} + (f)$. Then $\mathcal{Y} \geq 0$, $\mathcal{Y}$ satisfies (a), (b), (c), and $\mathcal{Y}$ is prime to $\mathfrak{p}_1, \ldots, \mathfrak{p}_s$ (otherwise $\mathcal{Y} - \mathfrak{p}_i \geq 0$ contradicts the choice of f). Now by (b), $\dim (\mathcal{Y} - \sum_{j \neq i} \mathfrak{O}_j) \geq 1$. In fact, it is exactly 1 by the argument used above for $\mathcal{K}$. Let $w_i \in \mathfrak{L}(\mathcal{Y} - \sum_{j \neq i} \mathfrak{O}_j)$, $w_i \neq 0$. Then $(w_i) + \mathcal{Y} - \sum_{j \neq i} \mathfrak{O}_j \geq 0$. Since the $\mathfrak{O}$'s are included in the $\mathfrak{p}$'s, $\mathfrak{O}_j$ must occur in (w_i), i.e. $w_i \equiv 0 \bmod \mathfrak{O}_j$, for $j \neq i$. But $w_i \not\equiv 0 \bmod \mathfrak{O}_i$ else $w_i \in \mathfrak{L}(\mathcal{Y} - \sum \mathfrak{O}_j) = 0$ by (c). By multiplying w_i by an element of k we can insure that $w_i \equiv 1 \bmod \mathfrak{O}_i$.

REMARK. The calculations used in [3] to deduce the Riemann hypothesis from Theorem 4.1 are all immediate from Lemma 3.4. Let k_0 be a finite field and let K_0 be a function field of dimension 1 over k_0. Let k be the algebraic closure of k_0 and let $K = k \otimes_{k_0} K_0$. Let $q = |k_0|$ and define $\Phi_n : K \to K$

by sending $a \otimes x \in k \otimes_{k_0} K_0$ to $a \otimes x^{q^n}$. Let $V_n = K$ with the usual left action of K, the right action of K being through Φ_n. Let ι_n be the class of V_n and δ the class of $K = KK/\widetilde{\mathfrak{S}}$. By Theorem 4.1, $\sigma((a\delta + b\iota_n)(a\delta + b\iota_n)') \geq 0$. As in [3], it follows that $\sigma(\iota_n)^2 \leq \sigma(\delta)\,\sigma(\iota_n \iota_n')$ using the obvious fact that $\sigma(\xi') = \sigma(\xi)$. By Lemma 3.4, the right side is $2gq^n$ since $|K : \Phi_n K| = q^n$. By Lemma 3.4 again, $f_{\mathfrak{p}}(V_n) = f_{\mathfrak{p}}(V_n K') = \ell(\mathfrak{O}_{\mathfrak{p}}/(\Phi_n(x) - x,\ \text{all } x \in \mathfrak{O}_{\mathfrak{p}}))$. To evaluate this, note that $\mathfrak{O}_{\mathfrak{p}}$ is unramified over $\mathfrak{O}_{\mathfrak{p}} = \mathfrak{O}_{\mathfrak{p}} \cap K_0$, the residue field of $\mathfrak{O}_{\mathfrak{p}}$ is a factor of $k \otimes_{k_0} \mathfrak{O}_{\mathfrak{p}_0}/\mathfrak{p}_0$ and Φ_n acts on this by $a \otimes x \mapsto a \otimes x^{q^n}$. Thus $\Phi_n(x) - x \in \mathfrak{p}$ for all $x \in \mathfrak{O}_{\mathfrak{p}}$ if and only if $\mathfrak{O}_{\mathfrak{p}_0}/\mathfrak{p}_0 \subset \mathbb{F}_{q^n}$, i.e. $\deg \mathfrak{p}_0$ divides n. In this case $f_{\mathfrak{p}} = 1$ since if $\pi \in \mathfrak{O}_{\mathfrak{p}_0}$ is prime then $\Phi_n(\pi) - \pi = \pi^{q^n} - \pi$ has has order 1 at $\mathfrak{p}$. Therefore $\sum_{\mathfrak{p}} f_{\mathfrak{p}}(V_n) = \sum_{\deg \mathfrak{p}_0 | n} \deg \mathfrak{p}_0 = N_n$ (the number of points on the curve rational over $\mathbb{F}_{q^n}$). Since $d'(V_n) = 1$ and $d''(V_n) = q^n$ we get $|\sigma(\iota_n)| = |1 + q^n - N_n| \leq 2gq^{n/2}$ which is equivalent to the Riemann hypothesis [3].

The calculations in [3] leading to a proof of Artin's conjecture for function fields can be done here in exactly the same way.

REFERENCES

[1] M. Eichler, Introduction to the theory of algebraic numbers and functions, Academic Press, New York and London (1966).

[2] P. Roquette, Arithmetischer Beweis der Riemannschen Vermutung in Kongruenzfunktionenkörpern beliebigen Geschlechts, J. Reine Angew. Math. 191, 199-252 (1953).

[3] A. Weil, Les Courbes algebriques et les varietes qui s'en deduisent, Hermann, Paris 1948.

THE STRUCTURE OF THE WITT RING AND QUOTIENTS OF ABELIAN GROUP RINGS

Roger Ware

Northwestern University

In 1937, Witt [9] introduced a commutative ring, now called the Witt ring, whose elements are isometry classes of anisotropic quadratic forms over a field of characteristic different from two. During the next thirty years very little was known about the structure of this ring. Then, in 1966, Pfister [7] proved a number of structure theorems for the Witt ring using his theory of multiplicative forms. Subsequently, Harrison [2] and Leicht and Lorenz [5] supplemented Pfister's results by completely determining the prime ideal structure of the Witt ring. The purpose of this article is to indicate how most of these results can be recovered using purely ring theoretic techniques applied to certain factor rings of integral abelian group rings. This stems from the fact that the Witt ring is itself a homomorphic image of a group ring $\mathbb{Z}[G]$ where G is a group of exponent two.

This lecture summarizes some joint work with M. Knebusch and A. Rosenberg [4].

<u>DEFINITION OF THE WITT RING</u>. In this section we briefly recall the definition of the Witt ring for a field F of characteristic not two.

A <u>quadratic</u> <u>F-space</u> is a pair (V,B) where V is a finite dimensional F-module and $B: V \times V \to F$ is a symmetric non-degenerate bilinear form.

If (V,B) and (V',B') are two spaces we write $(V,B) \cong (V',B')$ (or simply $V \cong V'$) to mean V is isometric to V'.

We write $V = V_1 \perp \cdots \perp V_r$ if $V = V_1 \oplus \cdots \oplus V_r$ with $B(V_i,V_j) = 0$ for $i \neq j$.

The space V is called <u>anisotropic</u> if $B(x,x) = 0$ implies $x = 0$. At the other extreme, V is called <u>hyperbolic</u> if $\dim_F V = 2n$ and V has an n-dimensional subspace U with $B(x,y) = 0$ for all x, y in U (equivalently, if V has a basis $x_1, \ldots, x_{2n}$ with $B(x_i,x_j) = 0$ for $i \neq j$ and $B(x_i,x_i) = (-1)^i$).

<u>Witt's decomposition</u> [9, Satz 5]. Let V be a quadratic F-space. Then there exists an anisotropic space V_a and a hyperbolic space V_h such that $V \cong V_a \perp V_h$. Moreover, the spaces V_a and V_h are unique up to isometry.

Thus if one wants to classify quadratic spaces over F it's enough to know the anisotropic ones.

Let $W(F)$ be the set of all isometry classes $[V]$ of anisotropic spaces over F. Then $W(F)$ becomes a commutative ring with 1, called the <u>Witt ring</u> of F, if addition and multiplication are defined by

$$[(V_1,B_1)] + [(V_2,B_2)] = [(V_1 \oplus V_2, B_1 + B_2)_a]$$

and

$$[(V_1,B_1)][(V_2,B_2)] = [(V_1 \otimes_F V_2, B_1 \otimes B_2)_a].$$

REMARKS:

(1) If $V = V_1 \perp \cdots \perp V_r$ then $[V] = [V_1] + \ldots + [V_r]$ in $W(F)$.

(2) V is hyperbolic if and only if $[V] = 0$ in $W(F)$.

(3) The additive inverse of $[(V,B)]$ is $[(V,-B)]$.

EXAMPLES:

(1) $W(\mathbb{R}) \cong \mathbb{Z}$ (Sylvester's law of inertia); $W(\mathbb{C}) \cong \mathbb{F}_2$, the field with two elements.

(2) If p is an odd prime, then $W(\mathbb{F}_p) \cong \mathbb{Z}/4\mathbb{Z}$ if $p \equiv 3 \pmod 4$ and $W(\mathbb{F}_p) \cong \mathbb{F}_2[G]$, G cylic of order 2, if $p \equiv 1 \pmod 4$.

(3) (Milnor [6]). As an additive group,

$$W(\mathbb{Q}) \cong \mathbb{Z} \oplus \mathbb{Z}/2\mathbb{Z} \oplus \coprod_p W(\mathbb{F}_p),$$

where the direct sum extends over all odd primes p.

PFISTER'S THEOREMS AND THE IDEAL THEORY OF $W(F)$. In his important paper [7], Pfister proved the following theorems about the structure of the Witt ring.

I. [7, Satz 10]. The order of any torsion element in the additive group of $W(F)$ is a power of two.

II. [7, Satz 11]. If $[V]$ is a zero divisor in $W(F)$ then $\dim_F V$ is even.

III. [7, Sätze 17, 22]. An element $[V]$ of $W(F)$ is nilpotent if and only if $[V]$ is a torsion element and $\dim_F V$ is even.

IV. [7, Sätze 16, 17]. Let s be the stufe of F (= least number s such that -1 is a sum of s squares). If $s < \infty$ then

(a) $W(F)$ is additively a torsion group of exponent $2s$.

(b) $W(F)$ has a unique prime ideal, consisting of all classes of even dimensional spaces.

Concerning the ideal theory of $W(F)$, Pfister also showed that the Jacobson radical of $W(F)$ is equal to the nil radical [7, Satz 23]. Harrison [2] and Leicht and Lorenz [5] proved that the non-maximal prime ideals of $W(F)$ correspond bijectively with the (total) orderings on the field F. Other results about the prime ideal theory of the Witt ring can be summarized as follows:

$W(F)$ has a unique maximal ideal $I(F)$ containing 2, $I(F)$ consists of all even dimensional forms, and $W(F)/I(F) \cong \mathbb{F}_2$. If $M \neq I(F)$ is a maximal ideal of $W(F)$, then $W(F)/M \cong \mathbb{F}_p$, p an odd prime, and M contains a unique non-maximal prime ideal P. A prime ideal P is non-maximal if and only if $W(F)/P \cong \mathbb{Z}$.

REMARKS: If F is an ordered field then one can topologize the set Ω of orderings on F by taking a subbase of open sets to be all sets of the form $W(a) = \{< \text{ in } \Omega \mid a < 0\}$, for $a \neq 0$ in F. Employing the correspondence between orderings and non-maximal prime ideals of $W(F)$ together with the above facts, Milnor [6] noted that $\mathrm{Spec}(W(F))$ is homeomorphic to the quotient space which is obtained from the cartesian product $\Omega \times \mathrm{Spac}(\mathbb{Z})$ by collapsing $\Omega \times \{2\}$ to a point. (See also [2,p. 13]).

PRESENTATION OF $W(F)$ AND QUOTIENTS OF ABELIAN GROUP RINGS. For a in $\dot{F} = F - 0$, let $\langle a \rangle$ denote the one-dimensional space F with form $(x,y) \to axy$ and let $\underline{a} \in \dot{F}/\dot{F}^2$ denote the square class of a. Then $\langle a \rangle$ is anisotropic, $\langle a \rangle \cong \langle b \rangle$ if and only if $\underline{a} = \underline{b}$, and $\langle a \rangle \otimes_F \langle b \rangle \cong \langle ab \rangle$. Hence the mapping $\dot{F} \to W(F)$ via $a \to [\langle a \rangle]$ induces a ring homomorphism $\varphi : \mathbb{Z}[G] \to W(F)$ where $G = \dot{F}/\dot{F}^2$ is the group of square classes. Since every quadratic form over F can be diagonalized, φ is surjective. Moreover, from [9, Satz 7] it follows that the kernel K of φ is generated as an ideal by $\underline{1} + \underline{-1}$ and all elements of the form $\underline{a} + \underline{b} - \underline{c} - \underline{d}$ with $\langle a \rangle \perp \langle b \rangle \cong \langle c \rangle \perp \langle d \rangle$. In particular, if $\psi : \mathbb{Z}[G] = \mathbb{Z}$ is any ring homomorphism then $\psi(K) = 2^n\mathbb{Z}$ or 0.

Now let G be an abelian p-group, p prime. Then $\mathbb{Z}[G]$ has a unique maximal ideal, $I_p(G)$, containing p; namely the kernel of the augmentation map followed by reduction mod p. Thus for the Witt ring (with $G = \dot{F}/\dot{F}^2$) we have $I(F) = I_2(G)/K$. Let L be the field generated over $\mathbb{Q}$ by all $\chi(g)$, g in G, where χ runs through all characters (i.e. homomorphisms) $\chi : G \to \mathbb{C}$, and let A be the integral closure of $\mathbb{Z}$ in L. Note that if G is a group of exponent two (e.g. $G = \dot{F}/\dot{F}^2$) then $A = \mathbb{Z}$. Then it is easy to see that the minimal prime ideals of $\mathbb{Z}[G]$ are the kernels of the induced maps $\psi_\chi : \mathbb{Z}[G] \to A$ and the maximal ideals of $\mathbb{Z}[G]$ are of the form $M_{\chi,\mathfrak{p}} = \psi_\chi^{-1}(\mathfrak{p})$ where $\mathfrak{p}$ is a non-zero prime ideal (i.e. maximal ideal) of A [4, Lemma 3.1].

Moreover, it turns out that every maximal ideal $M \neq I_p(G)$ of $\mathbb{Z}[G]$ contains a unique non-maximal (= minimal) prime ideal [4, Theorem 2.14].

With regard to Pfister's theorems one has the following:

THEOREM (cf. [4, Theorem 3.9]). Let G be an abelian p-group, K an ideal of $\mathbb{Z}[G]$, and $R = \mathbb{Z}[G]/K$. Then the following statements are equivalent:

(i) For every homomorphism $\psi : \mathbb{Z}[G] \to A$, $\psi(K) \cap \mathbb{Z} = p^n\mathbb{Z}$ or 0.

(ii) If $M \neq I_p(G)$ is a maximal ideal of $\mathbb{Z}[G]$ with $K \subset M$ then there exists a (unique) non-maximal prime ideal P with $K \subset P \subset M$.

(iii) $K \subset I_p(G)$ and the nil radical of R coincides with the torsion subgroup of $I_p(G)/K$.

(iv) The torsion subgroup of R is p-primary.

(v) $K \subset I_p(G)$ and all zero divisors of R lie in $I_p(G)/K$.

COROLLARY. Assume $R = \mathbb{Z}[G]/K$ satisfies the equivalent conditions of the theorem. Then R is a torsion group if and only if $I_p(G)/K$ is the only prime ideal of R.

COROLLARY. If R satisfies the conditions of the theorem then R is connected (i.e. has no idempotents except 0 and 1) and R has non-zero torsion if and only if $I_p(G)/K$ consists entirely of zero divisors.

REMARKS:

(1) The proof of the theorem requires only elementary techniques from commutative algebra and a minimal knowledge of abelian group theory. In fact, all the implications except (iii) $\Rightarrow$ (iv) are very straightforward. For a detailed proof, see [4].

(2) If G is any abelian torsion group, K an ideal of $\mathbb{Z}[G]$ and $R = \mathbb{Z}[G]/K$ then

(a) Every prime ideal of R is an intersection of maximal ideals. In particular, the nil radical of R equals the Jacobson radical.

(b) Every nilpotent element of R is a torsion element.

Statement (a) is a consequence of the fact that R is an integral extension of $\mathbb{Z}$ and (b) follows from the fact that $\mathbb{Q} \otimes_{\mathbb{Z}} R$ is a ring with no nilpotent elements and the torsion subgroup of any abelian group M is the kernel of the map $M \to \mathbb{Q} \otimes_{\mathbb{Z}} M$.

(3) Many other Witt-like rings have been defined in the literature. For example, Scharlau [8] and Belskii [1] have introduced Witt-Grothendieck and Witt rings for a profinite group which extend the notions of Witt-Grothendieck and Witt rings of quadratic forms over a field. These rings have the form $\mathbb{Z}[G]/K$ with G a group of exponent two and K an ideal which under every homomorphism of $\mathbb{Z}[G]$ to $\mathbb{Z}$ is mapped onto $2^n\mathbb{Z}$ or 0. If C is a connected semi-local ring the same is true for the Witt ring W(C) and Witt-Grothendieck ring K(C) of symmetric bilinear forms over C as defined by Knebusch [3], and also for the similarly defined rings W(C,J) and K(C,J) of hermitian forms over C with respect to some involution J [4].

Thus one can use the theorem to obtain unified proofs for the structure of all these rings.

REFERENCES

[1] A. A. Belskii, Cohomological Witt Rings, Math. USSR-Izvestija 2 (1968), pp. 1101-1115.

[2] D. K. Harrison, Witt rings, Lecture notes, Department of Mathematics, University of Kentucky, Lexington, Kentucky, 1970.

[3] M. Knebusch, Grothendieck and Witt ringe von nichtausgearteten symmetrischen Bilinearformen, Sitzber. Heidelberg Akad. Wiss. 1969/1970, pp. 93-157.

[4] M. Knebusch, A. Rosenberg, and R. Ware, Structure of Witt rings and quotients of abelian group rings, Amer. J. Math., (to appear).

[5] J. Leicht und F. Lorenz, Die Primideale des Wittschen Ringes, Invent. Math. 10 (1970), pp. 82-88.

[6] J. Milnor (with D. Husemoller), Symmetric bilinear forms, Lecture notes, Institute for Advanced Study, Princeton, 1971.

[7] A. Pfister, Quadratische Formen in Beliebigen Körpern, Invent. Math. 1 (1966), 116-132.

[8] W. Scharlau, Quadratische Formen und Galois-Cohomologie, Invent. Math. 4 (1967), pp. 238-264.

[9] E. Witt, Theorie der quadratische Formen in beliegigen Körpern, J. Reine Angew. Math. 176 (1937), pp. 31-44.

THE SCHUR SUBGROUP OF THE BRAUER GROUP(*)

Toshihiko Yamada

Queen's University

Let k be a field of characteristic 0. If A is a central simple algebra over k then [A] will denote the class of A in the Brauer group Br(k) of k. The Schur subgroup S(k) of Br(k) consists of those algebra classes [A] of Br(k) such that A is a simple component of the group algebra kG for some finite group G. The purpose of this paper is to survey recent results about the structure of the Schur subgroup S(k).

In §1 we will review the Brauer-Witt theorem and see that the Schur subgroup S(k) is nothing but the cyclotomic subgroup C(k). In §2 the Schur subgroup S(k) is characterized for the case k is a finite extension of the rational p-adic field $\mathbb{Q}_p$, where p is an odd prime number. In §3 several properties of elements of S(k) are stated. In §4 we characterize S(k) for some real cyclotomic extensions k of the rational field $\mathbb{Q}$. The case k is imaginary is dealt with in §5.

§1. THE BRAUER-WITT THEOREM. Throughout this section k denotes a field of characteristic 0. The Brauer group Br(k) is the inductive limit of the second cohomology groups $H^2(\mathcal{G}(L/k), L^x)$, where L ranges over all the finite Galois extensions of k and $\mathcal{G}(L/k)$ is the Galois group of L over k. If A is a central simple algebra over k there exists a finite Galois extension L of k which splits A. Then A is similar to a crossed product of L/k with a factor set $\alpha : \alpha(\sigma,\tau)^\rho\, \alpha(\rho,\sigma\tau) = \alpha(\rho,\sigma)\, \alpha(\rho\sigma,\tau)$, $\rho, \sigma, \tau \in \mathcal{G} = \mathcal{G}(L/k)$,

(*) This work was done while the author was a visiting associate professor at Queen's University in 1971/72.

which is nothing but a 2-cocycle of $\mathfrak{g}$ with values in $L^{\times}$. By this correspondence the subgroup of $Br(k)$ consisting of those algebra classes $[A]$ which are split by L is identified with $H^2(\mathfrak{g}, L^{\times})$. Let K be an extension of k. Denote by Res the restriction homomorphism of $Br(k)$ into $Br(K)$. It follows from [7, V, §4, Satz 1] that the image $Res[A]$ of $[A] \in Br(k)$ by Res is $[A \otimes_k K]$.

A <u>cyclotomic algebra</u> (Kreisalgebra) over k is a crossed product

$$B = (\beta, k(\zeta)/k) = \sum_{\sigma \in \mathfrak{g}} k(\zeta) u_{\sigma} \quad \text{(direct sum)},$$

$$u_{\sigma} x = x^{\sigma} u_{\sigma} \ (x \in k(\zeta)), \ u_{\sigma} u_{\tau} = \beta(\sigma, \tau) u_{\sigma\tau},$$

where ζ is a root of unity, $\mathfrak{g}$ is the Galois group of $k(\zeta)$ over k, and β is a factor set of $k(\zeta)/k$ whose values are roots of unity in $k(\zeta)$. The <u>cyclotomic subgroup</u> $C(k)$ of the Brauer group $Br(k)$ of k consists of those algebra classes which contain a cyclotomic algebra over k.

LEMMA 1.1. <u>Every cyclotomic algebra</u> B <u>over</u> k <u>is</u> k-<u>isomorphic to a simple component of</u> kG, <u>where</u> G <u>is a certain finite subgroup of the multiplicative group</u> $B^{\times}$. <u>In particular, the class</u> $[B]$ <u>belongs to</u> $S(k)$.

PROOF: Let B be a cyclotomic algebra over k:

$$B = (\beta, k(\zeta')/k) = \sum_{\sigma \in \mathfrak{g}} k(\zeta') u_{\sigma}$$

$$u_{\sigma} x = x^{\sigma} u_{\sigma} (x \in k(\zeta')), \ u_{\sigma} u_{\tau} = \beta(\sigma, \tau) u_{\sigma\tau}$$

where ζ' is a root of unity, $\mathfrak{g} = \mathfrak{g}(k(\zeta')/k)$, and β is a factor set whose values are roots of unity in $k(\zeta')$. It is clear that ζ' and the values of β generate a finite cyclic group $\langle \zeta \rangle$ in $k(\zeta')^{\times}$ and $k(\zeta') = k(\zeta)$, where ζ is some root of unity. The Galois group $\mathfrak{g}$ can be regarded as an automorphism group of the cyclic group $\langle \zeta \rangle$ and the values of the factor set β belong to $\langle \zeta \rangle$. By the theory of group extension (cf. for instance, [16, IV, §6]) we see easily that ζ and $u_\sigma (\sigma \in \mathfrak{g})$ generate a finite subgroup G in B^x. That is, G has a normal cyclic subgroup $\langle \zeta \rangle$, the factor group $G/\langle \zeta \rangle$ is isomorphic to $\mathfrak{g}$, and β is exactly a factor set of the extension G of $\langle \zeta \rangle$ by $\mathfrak{g}$. Since G spans B with coefficients in k, there exists a k-algebra-homomorphism of kG onto B, and so B is k-isomorphic to a simple component of kG.

Now we will review the Brauer-Witt theorem. Let L be a field of characteristic 0. If G is a finite group and χ is its absolutely irreducible character with $L(\chi) = L$, then $A(\chi,L)$ denotes the simple component of the group algebra LG which corresponds to χ, i.e. $\chi(A(\chi,L)) \neq 0$.

BRAUER-WITT THEOREM. Let G be a finite group of exponent n and χ an absolutely irreducible character of G. Let p be a prime number.

(i) Let k be a field of characteristic 0 with $k(\chi) = k$. Let L be the subfield of $k(\zeta_n)$ over k such that $[k(\zeta_n) : L]$ is a p-power and $[L : k] \not\equiv 0 \pmod p$, where ζ_n is a primitive nth root of unity. Then there exists a subgroup F of G which is L-elementary with respect to p and an absolutely irreducible character ξ of F with $L(\xi) = L$ such that $(\chi | F, \xi) \not\equiv 0 \pmod p$ and that the simple component $A(\xi, L)$ is a cyclotomic algebra over L; moreover, the order of the class $[A(\xi,L)]$ in $Br(L)$ is a p-power.

(ii) If K is a field of characteristic 0 with $K(\chi) = K$, then for every subgroup H of G and for every absolutely irreducible character ψ of H such that $K(\psi) = K$ and $(\chi | H, \psi) \not\equiv 0 \pmod p$, the p-parts of the classes $[A(\chi, K)]$ and $[A(\psi, K)]$ in $Br(K)$ are the same.

COROLLARY 1.2. (Fontaine [10, Proposition 8.1]). Keeping the notation of the

<u>Brauer-Witt theorem, the class</u> $[A(\chi, k)]$ <u>belongs to the cyclotomic subgroup</u> $C(k)$.

PROOF: Let $[k(\zeta_n):L] = p^s$ and $[L:k] = t \not\equiv 0 \pmod p$. Let t' be an integer such that $tt' \equiv 1 \pmod{p^s}$. As $A(\chi, L) = A(\chi, k) \otimes_k L$, it follows that $[A(\chi, L)] = \mathrm{Res}[A(\chi, k)]$. Denote by Cor the corestriction homomorphism: $Br(L) \to Br(k)$. Then it is readily checked that $C(L)$ is mapped into $C(k)$ by Cor. From the Brauer-Witt theorem it follows that the p-part $[A(\chi, L)]_p$ of $[A(\chi, L)]$ equals $[A(\xi, L)]$, which is in $C(L)$. Since $k(\zeta_n)$ is a splitting field of $A(\chi, k)$, the order of $[A(\chi, k)]_p$ in $Br(k)$ divides p^s. As $[A(\xi, L) = [A(\chi, L)]_p = \mathrm{Res}([A(\chi, k)]_p)$, we have $(\mathrm{Cor}[A(\xi, L)])^{t''} =$ $(\mathrm{Cor} \cdot \mathrm{Res}([A(\chi, k)]_p))^{t'} = ([A(\chi, k)]_p)^{tt'} = [A(\chi, k)]_p$. Hence $[A(\chi, k)]_p$ is in $C(k)$. As p is an arbitrary prime, $[A(\chi, k)]$ is in $C(k)$.

REMARK. The Brauer-Witt theorem was first proved by R. Brauer [5], [6], in which the assertion (II) was given in terms of the Schur index. Subsequently, E. Witt [12] obtained the theorem by another method.

From Lemma 1.1 and Corollary 1.2 we get

THEOREM 1.3. <u>The Schur subgroup</u> $S(k)$ <u>and the cyclotomic subgroup</u> $C(k)$ <u>are the same subgroup of the Brauer group</u> $Br(k)$, i.e. $S(k) = C(k)$.

Thus in order to characterize the Schur subgroup $S(k)$ in $Br(k)$ we need only characterize the cyclotomic subgroup $C(k)$ in $Br(k)$ and every property of elements of $C(k)$ is indeed a property of elements of $S(k)$, and vice versa. A cyclotomic algebra is much easier to study than an abritrary simple component of a group algebra, but there are some cases in which characterizing $S(k)$ is easier than characterizing $C(k)$.

§2. THE STRUCTURE OF $S(k)$ FOR $k \supset \mathbb{Q}_p$. Let p be a prime number. Consider a p-adic cyclotomic algebra B:

$$B = (\beta(\sigma,\tau), L/k) = \sum_{\sigma} L u_{\sigma}$$

$$u_{\sigma} x = x^{\sigma} u_{\sigma} \ (x \in L), \quad u_{\sigma} u_{\tau} = \beta(\sigma,\tau) u_{\sigma\tau}$$

with the following properties:

(i) L is a cyclotomic field $\mathbb{Q}_p(\zeta_n)$ over the rational p-adic field, where ζ_n is a primitive nth root of unity for a positive integer n,

(ii) k is a subfield of L over $\mathbb{Q}_p$ and the center of B,

(iii) $\beta(\sigma,\tau)$ is a factor set of $\mathfrak{G}(L/k)$ with values in the group $W(L)$ of all the roots of unity in L.

Let $n = p^h t$, $(p,t) = 1$, $N_{k/\mathbb{Q}_p}(\mathfrak{P}) = y$, where $\mathfrak{P}$ is the prime of k and $N_{k/\mathbb{Q}_p}$ is the norm of k over $\mathbb{Q}_p$. Let e be the ramification index of L/k and f the residue class degree of L/k. Then $W(L) = \langle \zeta_{y^f-1} \rangle \times \langle \zeta_{p^h} \rangle$. For any σ, τ of $\mathfrak{G}(L/k)$, let $\beta(\sigma,\tau) = \alpha(\sigma,\tau)\gamma(\sigma,\tau)$, $\alpha(\sigma,\tau) \in \langle \zeta_{y^f-1} \rangle$, $\gamma(\sigma,\tau) \in \langle \zeta_{p^h} \rangle$. If p is an odd prime number, the multiplicative group $\mathbb{Z} \bmod^{x} p^h$ of integers modulo p^h is cyclic, and the inertia group T of L/k is isomorphic to a subgroup of $\mathbb{Z} \bmod^{x} p^h$. Let ω be a generator of the cyclic group $T : T = \langle \omega \rangle$, $\omega^e = 1$. Let η be a Frobenius automorphism of L/k. Then $\mathfrak{G}(L/k) = \langle \omega, \eta \rangle$.

THEOREM 2.1. ([13, Theorem 3, p. 302]). <u>Notation and assumption being as above, let</u> c <u>be the index of tame ramification of</u> $\mathfrak{P}$ <u>over</u> p, i.e. $p = \pi^{cp^{\lambda}}$, $(c,p) = 1$, <u>for a certain integer</u> λ. <u>Then the number</u>

$$\Delta = (\alpha(\omega,\eta)/\alpha(\eta,\omega))^{e/(y-1)} \alpha(\omega,\omega)\, \alpha(\omega^2,\omega) \cdots \alpha(\omega^{e-1},\omega)$$

<u>belongs to</u> k, <u>so that we can write it as</u> $\Delta = \zeta_{y-1}^{v}$ <u>for a certain integer</u> v. <u>The index of the cyclotomic algebra</u> $B = (\beta(\sigma,\tau), L/k)$ <u>is equal to</u>

$$\frac{(p-1)/c}{(v, (p-1)/c)} .$$

Conversely, for any integer z we can construct a cyclotomic algebra over k with Hasse invariant $z/\frac{p-1}{c}$ (cf. [13, pp. 307-308]). Hence we have

THEOREM 2.2. (Fontaine [10] and Yamada [13]). *Let* p *be an odd prime number. Let* k *be a cyclotomic extension of the rational* p*-adic field* $\mathbb{Q}_p$. *Denote by* c *the tame ramification index of* $k/\mathbb{Q}_p$. *Then a class* $[A]$ *of* $Br(k)$ *belongs to the Schur subgroup* $S(k)$ *of* k *if and only if the Hasse invariant of* A *is of the form* $z/\frac{p-1}{c}$, $z \in Z$.

From this theorem we can easily determine the Schur subgroup $S(K)$ for any finite extension K of $\mathbb{Q}_p$, $p \neq 2$ (cf. [13, Theorem 2]).

§3. SOME PROPERTIES OF ELEMENTS OF $S(k)$.

LEMMA 3.1. *Let* K *be a field. If a cyclotomic algebra* B *over* k *has exponent* m *in* $Br(k)$ *then the* mth *roots of unity lie in the center* k.

PROOF: See Janusz [11].

THEOREM 3.2. (Benard and Schacher [4]). *Let* k *be a finite abelian extension of the rational field* $\mathbb{Q}$. *Let* $[A] \in S(k)$ *and suppose* A *has index* m. *Then* $\zeta_m \in k$. *Let* p *be a rational prime. If* $\mathfrak{P}$ *is a prime of* k *dividing* p *and* $\theta \in \mathcal{G}(k/\mathbb{Q})$ *with* $\zeta_m^\theta = \zeta_m^b$, *then* $inv_{\mathfrak{P}}(A) \equiv b\, inv_{\mathfrak{P}^\theta}(A) \pmod 1$.

PROOF: The first statement follows from Theorem 1.3 and Lemma 3.1. Let B be a cyclotomic algebra over k with $[B] = [A]$ in $Br(k) : B = (\beta, k(\zeta)/k) = \sum_{\sigma \in \mathcal{G}} k(\zeta)u_\sigma$, $u_\sigma x = x^\sigma u_\sigma (x \in k(\zeta))$, $u_\sigma u_\tau = \beta(\sigma, \tau) u_{\sigma\tau}$, where ζ is a root of unity and the values of β are roots of unity in $k(\zeta)$. The automorphism θ

of $k/\mathbb{Q}$ can be extended to an automorphism of $k(\zeta)/\mathbb{Q}$, which is also denoted by θ. Set $\beta^{\theta}(\sigma,\tau) = (\beta(\sigma,\tau))^{\theta}$ for $\sigma,\tau \in \mathcal{G}(k(\zeta)/k)$. Then it is easily checked that β^{θ} is a factor set of $k(\zeta)/k$. Let B^{θ} be the cyclotomic algebra with the factor set β^{θ}: $B^{\theta} = (\beta^{\theta}, k(\zeta)/k) = \sum_{\sigma \in \mathcal{G}} k(\zeta)v_{\sigma}, v_{\sigma}x = x^{\sigma}v_{\sigma}(x \in k(\zeta)), v_{\sigma}v_{\tau} = \beta^{\theta}(\sigma,\tau)v_{\sigma\tau}$. Now map B onto B^{θ} by $\sum_{\sigma} x_{\sigma}u_{\sigma} \to \sum_{\sigma} x_{\sigma}^{\theta}v_{\sigma}$ $(x_{\sigma} \in k)$. As is easily seen, this is a $\mathbb{Q}$-algebra isomorphism and $\mathrm{inv}_{\mathfrak{P}}(B) = \mathrm{inv}_{\mathfrak{P}^{\theta}}(B^{\theta})$. Let the roots of unity in $k(\zeta)$ generate a cyclic group $\langle\zeta_n\rangle$ of nth roots of unity. Then m divides n. If $\zeta_n^{\theta} = \zeta_n^{t}$ for some integer t then $\zeta_m^{\theta} = \zeta_m^{t}$. Hence $t \equiv b \pmod{m}$, and so $\mathrm{inv}_{\mathfrak{P}}(B) = \mathrm{inv}_{\mathfrak{P}^{\theta}}(B^{\theta}) \equiv t\,\mathrm{inv}_{\mathfrak{P}^{\theta}}(B) \equiv b\,\mathrm{inv}_{\mathfrak{P}^{\theta}}(B) \pmod{1}$.

COROLLARY 3.3. (Benard [3]). If $\mathfrak{P}_1$, and $\mathfrak{P}_2$ are primes of k dividing p then $A \otimes_k k_{\mathfrak{P}_1}$ and $A \otimes_k k_{\mathfrak{P}_2}$ have the same index.

COROLLARY 3.4. (The Brauer-Speiser Theorem). Let χ be a real-valued irreducible character. Then the Schur index $m_{\mathbb{Q}}(\chi)$ of χ over $\mathbb{Q}$ is at most 2. If $\chi(1)$ is odd, then $m_{\mathbb{Q}}(\chi) = 1$.

Whenever a central simple algebra A has invariants which are distributed in the way described in Theorem 3.2 for all primes p, we say that A has uniformly distributed invariants.

§4. $S(k)$ FOR A REAL FIELD k. Let k be a real subfield of a cyclotomic field over $\mathbb{Q}$. The Brauer-Speiser theorem implies that every algebra class $[A]$ of $S(k)$ has Hasse invariant 0 or 1/2 at any prime $\mathfrak{P}$ of K. Let p be an arbitrary rational prime. From Theorem 3.2 it follows that A has the same Hasse invariant at all the primes $\mathfrak{P}$ of k dividing p.

DEFINITION 4.1. Let k be a real subfield of a cyclotomic field over $\mathbb{Q}$. Then k is called to have the property (M) if the Schur subgroup $S(k)$ consists of all those algebra classes that satisfy the above two conditions.

THEOREM 4.2. (Benard [2] and Fields [8]). <u>The rational field</u> $\mathbb{Q}$ <u>has the property</u> (M).

PROOF: Let p be an odd prime number. Consider the cyclic algebra $B_p = (-1, \mathbb{Q}(\zeta_p)/\mathbb{Q}, \sigma)$, $\langle \sigma \rangle = \mathfrak{G}(\mathbb{Q}(\zeta_p)/\mathbb{Q})$, which is a cyclotomic algebra. As is easily seen, B_p has Hasse invariant $1/2$ at p and ∞ and zero elsewhere. Denote by B_2 the quaternion algebra $(-1, \mathbb{Q}(\zeta_4)/(\mathbb{Q}, \sigma)$, $\zeta_4^{\sigma} = \zeta_4^{-1}$, which is a cyclotomic algebra. Then B_2 has Hasse invariant $1/2$ at 2 and ∞ and zero elsewhere. By taking tensor products, we see that every quaternion algebra central over $\mathbb{Q}$ appears in $S(\mathbb{Q})$.

COROLLARY 4.3. <u>Let</u> k <u>be a subfield of a cyclotomic field over</u> $\mathbb{Q}$. <u>If the degree of</u> k <u>over</u> $\mathbb{Q}$ <u>is odd then</u> k <u>is real and has the property</u> (M).

THEOREM 4.4. (Yamada [14]). <u>Let</u> q <u>be a prime number and</u> c <u>a positive integer. Let</u> k <u>be the maximal real subfield of the cyclotomic field</u> $\mathbb{Q}(\zeta_{q^c})$, <u>where</u> ζ_{q^c} <u>is a primitive</u> q^c th <u>root of unity</u>, i.e. $k = \mathbb{Q}(\zeta_{q^c} + \zeta_{q^c}^{-1})$. <u>Then</u> k <u>has the property</u> (M).

PROOF: See [14].

THEOREM 4.5. (Yamada [15]). <u>Let</u> q <u>and</u> q' <u>be distinct prime numbers. Then the following real quadratic fields have the property</u> (M).

(i) $\mathbb{Q}(\sqrt{q})$, <u>where</u> $q = 2$ <u>or</u> $q \equiv 1 \pmod 4$,

(ii) $\mathbb{Q}(\sqrt{qq'})$, <u>where</u> $q = 2$ <u>or</u> $q \equiv 1 \pmod 4$, <u>and</u> $q' \equiv 1 \pmod 4$ <u>such that the Legendre symbol</u> $\left(\frac{q}{q'}\right)$ <u>equals</u> -1.

To prove Theorem 4.5 we need a remark. Let k be a real quadratic field. For each rational prime p, q_p denotes the number of the primes of k dividing p. Let $\{p_1, p_2, \dots, p_n\}$ be any set of distinct rational primes such that the number of those p_i for which $q_{p_i} = 1$, is even. Denote by $\Omega(p_1, p_2, \dots, p_n)$ the algebra class of Br(k) which has Hasse invariant 1/2 at any $\mathfrak{P}$ of k lying above $p_1, p_2, \dots, p_n$ and zero elsewhere. Then it is easily seen that all those classes of Br(k) satisfying the two conditions of Definition 4.1 are exactly all those $\Omega(p_1, p_2, \dots, p_n)$, in which $\{p_1, p_2, \dots, p_n\}$ is a set of rational primes satisying the above condition. Thus in order to prove Theorem 4.5 it suffices to show that for any rational prime p with $q_p = 2$ the algebra class $\Omega(p)$ belongs to S(k) and that for any pair of distinct primes $\{p, p'\}$ with $q_p = q_{p'} = 1$ the class $\Omega(p, p')$ belongs to S(k), where k is a real quadratic field mentioned in Theorem 4.5.

LEMMA 4.6. Let k be a real quadratic field. Let p be a rational prime with $q_p = 2$ in k. Then the algebra class $\Omega(p)$ is in S(k).

PROOF: This follows easily from Theorem 4.2.

PROOF OF THEOREM 4.5: We know already that the field $\mathbb{Q}(\sqrt{2})$ has the property (M), because $\mathbb{Q}(\sqrt{2})$ is the maximal real subfield of $\mathbb{Q}(\zeta_8)$. Let $k = \mathbb{Q}(\sqrt{q})$, where q is a prime number with $q \equiv 1 \pmod 4$. By virtue of Lemma 4.6 and the remark preceding it, it suffices to prove that for any prime number p which is inertial in $k/\mathbb{Q}$ the class $\Omega(q, p)$ of Br(k) belongs to S(k). Let B be a cyclotomic algebra over k such that $\mathrm{inv}_{\mathfrak{Q}} B \equiv 1/2$ for the prime $\mathfrak{Q}$ of k dividing q ($q = \mathfrak{Q}^2$ in k) and that $\mathrm{inv}_{\ell} B \equiv 0$ for any finite prime ℓ of k dividing neither q nor p. Let $\infty = \mathfrak{P}_\infty \mathfrak{P}'_\infty$ be the decomposition of the rational infinite prime ∞ in k. As the class [B] is in C(k) = S(k), we have $\mathrm{inv}_{\mathfrak{P}_\infty} B \equiv \mathrm{inv}_{\mathfrak{P}'_\infty} B$. From the Hasse's sum theorem it follows that

$\mathrm{inv}_{\mathfrak{P}} B \equiv 1/2$, where $\mathfrak{P}$ is the prime of k lying above $p (p = \mathfrak{P}$ in $k)$. By Lemma 4.6 the class $\Omega(\infty)$ is in $S(k)$. Consequently, in order to prove $\Omega(q,p) \in S(k)$ it is enough to construct a cyclotomic algebra B with $\mathrm{inv}_{\mathfrak{P}} B \equiv 1/2$ and $\mathrm{inv}_{\ell} B \equiv 0$ for any finite prime ℓ of k dividing neither q nor p. Note that k is a subfield of $Q(\zeta_q)$.

(a) Assume first that $p \neq 2$. Set $K = Q(\zeta_q, \zeta_p)$. Then $K = Q(\zeta_q) \cdot k(\zeta_p)$ and $Q(\zeta_q) \cap k(\zeta_p) = k$. We have $\mathcal{G}(K/Q(\zeta_q)) = \langle\varphi\rangle$, $\mathcal{G}(K/k(\zeta_p)) = \langle\xi\rangle$, $\mathcal{G}(K/k) = \langle\varphi\rangle \times \langle\xi\rangle$. Consider the following cyclotomic algebra B_p:

$$B_p = (\varepsilon, K/k) = \sum_{i=0}^{p-2} \sum_{j=0}^{(q-1)/2-1} K u_\varphi^i u_\xi^j \quad \text{(direct sum)}$$

$$u_\varphi^i u_\xi^j x = x^{\varphi^i \xi^j} u_\varphi^i u_\xi^j \quad (x \in K)$$

$$u_\varphi u_\xi = -u_\xi u_\varphi, \quad u_\varphi^{p-1} = u_\xi^{(q-1)/2} = 1$$

$$(u_\varphi^i u_\xi^j)(u_\varphi^{i'} u_\xi^{j'}) = \varepsilon(\varphi^i \xi^j, \varphi^{i'} \xi^{j'}) u_\varphi^{i+i'-\delta_\varphi(p-1)} u_\xi^{j+j'-\delta_\xi(q-1)/2}$$

$$0 \leq i, i' \leq p-2, \quad 0 \leq j, j' \leq (q-1)/2 - 1.$$

$$\delta_\varphi = \begin{cases} 0, & i + i' < p - 1, \\ 1, & i + i' \geq p - 1, \end{cases} \qquad \delta_\xi = \begin{cases} 0, & j + j' < (q-1)/2, \\ 1, & j + j' \geq (q-1)/2. \end{cases}$$

For any finite prime ℓ of k we have $B_p \otimes_k k_\ell \sim 1$, because the values of the factor set ε are ± 1 and ℓ is not ramified in K/k. Using the formula of Theorem 2.1 we see that the $\mathfrak{P}$-index of B_p is two, as desired.

(b) Assume next that $p = 2$ is inertial in $k = Q(\sqrt{q})$. Then $q \equiv 5 \pmod 8$. Set $K = Q(\zeta_q, \zeta_4)$. Then $K = Q(\zeta_q) \cdot k(\zeta_4)$ and $Q(\zeta_q) \cap k(\zeta_4) = k$. Hence $\mathcal{G}(K/Q(\zeta_q)) = \langle\varphi\rangle$, $\mathcal{G}(K/k(\zeta_4)) = \langle\xi\rangle$, $\mathcal{G}(K/k) = \langle\varphi\rangle \times \langle\xi\rangle$. Consider the following cyclotomic algebra B_2:

$$B_2 = (\varepsilon, K/k) = \sum_{i=0}^{1} \sum_{j=0}^{(q-1)/2-1} K u_\varphi^i u_\xi^j$$

$$u_\xi u_\varphi = \zeta_4 u_\varphi u_\xi, \; u_\varphi^2 = 1, \; u_\xi^{(q-1)/2} = \zeta_4$$

$$(u_\varphi^i u_\xi^j)(u_\varphi^{i'} u_\xi^{j'}) = \epsilon(\varphi^i \xi^j, \varphi^{i'} \xi^{j'})\, u_\varphi^{i+i'-2\delta_\varphi} u_\xi^{j+j'-\delta_\xi (q-1)/2}$$

$$0 \le i, i' \le 1, \; 0 \le j, j' \le (q-1)/2 - 1,$$

$$\delta_\varphi = \begin{cases} 0, & i + i' < 2, \\ 1, & i + i' \ge 2, \end{cases} \qquad \delta_\xi = \begin{cases} 0, & j + j' < (q-1)/2, \\ 1, & j + j' \ge (q-1)/2. \end{cases}$$

By the same argument as before we conclude that the $\mathfrak{Q}$-index of B_2 equals 2 and for any finite prime ℓ of k the ℓ-index of B_2 equals 1. Thus we have shown that $k = Q(\sqrt{q})$, $q \equiv 1 \pmod 4$ has the property (M).

Let $k = Q(\sqrt{2q})$, where q is a prime number such that $q \equiv 1 \pmod 4$ and $\left(\frac{2}{q}\right) = -1$, i.e. $q \equiv 5 \pmod 8$. Let $\mathfrak{Q}$ (resp. $\mathfrak{D}$) be the prime of k lying above q (resp. 2), i.e., $q = \mathfrak{Q}^2$ (resp. $2 = \mathfrak{D}^2$). First we will construct a cyclotomic algebra $B_{q,2}$ over k such that $\mathrm{inv}_{\mathfrak{Q}} B_{q,2} \equiv \frac{1}{2}$ and $\mathrm{inv}_\ell (B_{q,2}) \equiv 0$ for any finite prime ℓ of k not dividing q nor 2. The Hasse's sum theorem implies that $\mathrm{inv}_{\mathfrak{D}} B_{q,2} \equiv \frac{1}{2}$. As the class $\Omega(\infty)$ is in $S(k)$ it follows that the class $\Omega(q,2)$ is in $S(k)$. Let p be a rational prime which is inertial in k and $\mathfrak{P}$ the prime of k lying above p, i.e. $p = \mathfrak{P}$. We will construct a cyclotomic algebra B_p over k such that $\mathrm{inv}_{\mathfrak{P}} (B_p) \equiv \frac{1}{2}$ and $\mathrm{inv}_\ell (B_p) \equiv 0$ for any finite prime ℓ of k other than $\mathfrak{P}, \mathfrak{Q}, \mathfrak{D}$. Again by the Hasse's sum theorem, either $\mathrm{inv}_{\mathfrak{Q}} (B_p) \equiv \frac{1}{2}$ and $\mathrm{inv}_{\mathfrak{D}} (B_p) \equiv 0$, or $\mathrm{inv}_{\mathfrak{Q}} (B_p) \equiv 0$ and $\mathrm{inv}_{\mathfrak{D}} (B_p) \equiv \frac{1}{2}$. This implies that one of the classes $\Omega(p,q)$ and $\Omega(p,2)$ is in $S(k)$. From this and the fact that $\Omega(q,2)$ is in $S(k)$, we conclude that for any rational primes p_1 and p_2 which are ramified or inertial in k, the class $\Omega(p_1,p_2)$ is in $S(k)$.

(i) Construction of $B_{q,2}$. Set $K = Q(\zeta_8, \zeta_q)$. Then $\mathcal{G}(K/Q) = \langle \iota \rangle \times \langle \theta \rangle \times \langle \varphi \rangle$, where $\zeta_8^\iota = \zeta_8^{-1}$, $\zeta_q^\iota = \zeta_q$, $\zeta_8^\theta = \zeta_8^5$, $\zeta_q^\theta = \zeta_q$, $\zeta_8^\varphi = \zeta_8$, $\zeta_q^\varphi = \zeta_q^r$, and r is a primitive root modulo q. Set $\psi = \varphi\theta$. As is easily seen, $\mathcal{G}(K/k) = \langle \psi \rangle \times \langle \iota \rangle$ and

$$\zeta_q^{\psi} = \zeta_q^{r}, \ \zeta_8^{\psi} = \zeta_8^{5}, \ \zeta_q^{\iota} = \zeta_q, \ \zeta_8^{\iota} = \zeta_8^{-1}. \tag{1}$$

Consider the following cyclotomic algebra $B_{q,2}$:

$$B_{q,2} = (\epsilon, K/k) = \sum_{i=0}^{q-2} \sum_{j=0}^{1} K u_{\psi}^{i} u_{\iota}^{j},$$

$$u_{\psi}^{i} u_{\iota}^{j} x = x^{\psi^i \iota^j} u_{\psi}^{i} u_{\iota}^{j} \qquad (x \in K)$$

$$u_{\iota} u_{\psi} = \zeta_8 u_{\psi} u_{\iota}, \ u_{\psi}^{q-1} = \zeta_4^{(q-1)/4}, \ u_{\iota}^2 = 1,$$

$$(u_{\psi}^{i} u_{\iota}^{j})(u_{\psi}^{i'} u_{\iota}^{j'}) = \epsilon(\psi^i \iota^j, \psi^{i'} \iota^{j'}) u_{\psi}^{i+i'-\delta_{\psi}(q-1)} u_{\iota}^{j+j'-2\delta_{\iota}}$$

$$0 \le i, i' \le q-2, \qquad 0 \le j, j' \le 1,$$

$$\delta_{\psi} = \begin{cases} 0, & i+i' < q-1, \\ 1, & i+i' \ge q-1, \end{cases} \qquad \delta_{\iota} = \begin{cases} 0, & j+j' < 2, \\ 1, & j+j' \ge 2. \end{cases}$$

If ℓ is any finite prime of k dividing neither q nor 2, then the ℓ-index of $B_{q,2}$ equals 1, because ℓ is unramified in K/k. Using the formula of Theorem 2.1 we can prove that the index of the q-adic cyclotomic algebra $B_{q,2} \otimes_k k_{\mathfrak{Q}}$ is 2.

(ii) Construction of B_p. Let p be a prime number which is inertial in k. Set $K = Q(\zeta_8, \zeta_q, \zeta_p)$. Then $K = \mathbb{Q}(\zeta_8, \zeta_q) \cdot k(\zeta_p)$, $\mathbb{Q}(\zeta_8, \zeta_p) \cap k(\zeta_p) = k$, $\mathcal{G}(K/k) = \langle \xi \rangle \times \langle \psi \rangle \times \langle \iota \rangle$, $\mathcal{G}(K/\mathbb{Q}(\zeta_8, \zeta_q)) = \langle \xi \rangle$, $\mathcal{G}(K/k(\zeta_p)) = \langle \psi \rangle \times \langle \iota \rangle$, where the automorphisms ψ and ι are defined by (1). Consider the following cyclotomic algebra B_p over k:

$$B_p = (\epsilon, K/k) = \sum_{i=0}^{p-2} \sum_{j=0}^{q-2} \sum_{h=0}^{1} K u_{\xi}^{i} u_{\psi}^{j} u_{\iota}^{h}$$

$$u_{\xi}^{i} u_{\psi}^{j} u_{\iota}^{h} x = x^{\xi^i \psi^j \iota^h} u_{\xi}^{i} u_{\psi}^{j} u_{\iota}^{h}$$

$$u_\xi^{p-1} = \zeta_8^{(p-1)/2}, \quad u_\psi^{q-1} = \zeta_4^{(q-1)/4}, \quad u_\iota = 1$$

$$u_\iota u_\psi = \zeta_8 u_\psi u_\iota, \quad u_\psi u_\xi = \zeta_4 u_\xi u_\psi, \quad u_\xi u_\iota = \zeta_8 u_\iota u_\xi.$$

If ℓ is any finite prime of k dividing neither 2, q nor p, then $B_p \otimes_k k_\ell \sim 1$. Let $\mathfrak{P}$ be the prime of k lying above p, i.e. $p = \mathfrak{P}$. By Theorem 2.1 we can prove that the $\mathfrak{P}$-index of B_p is 2. Thus the field $k = \mathbb{Q}(\sqrt{2q})$, q 5 (mod 8) has the property (M). For the case of $k = \mathbb{Q}(\sqrt{qq'})$, see [15].

Finally we remark that the real quadratic field $\mathbb{Q}(\sqrt{m})$ in which m is a square free positive integer divisible by a prime $q \equiv 3 \pmod 4$ does not have the property (M). This follows from Theorem 2.2.

§5. S(k) FOR AN IMAGINARY FIELD k. Let k be an imaginary subfield of a cyclotomic field over $\mathbb{Q}$. Let k_0 be the maximal real subfield of k and ι the automorphism of k/k_0 of order 2. Let G be a finite group and let $*$ denote the involution of kG defined by

$$* : \sum a_g g \to \sum a_g^\iota g^{-1} \qquad (a_g \in k).$$

Let χ be an irreducible character of G with $k(\chi) = k$. The simple component $A(\chi, k)$ of kG corresponding to χ is $kGe(\chi)$, $e(\chi) = \chi(1)\,|G|^{-1} \sum \chi(g^{-1})\,g$. As is easily seen, $*(e(\chi)) = e(\chi)$, and so the involution $*$ of kG induces an involution of $A(\chi, k)$ of the second kind. From [1, Chapter X, Theorem 21] it follows that if $A(\chi, k)$ has index 2 then $A(\chi, k) \sim D \otimes_{k_0} k$ for some quaternion division algebra D central over k_0. From this combined with Theorem 3.2 and Theorem 4.2 we have

THEOREM 5.1. (Benard-Schacher [4]). Let k be an imaginary quadratic field with $\sqrt{-1} \notin k$ and $\sqrt{-3} \notin k$. Then $S(k)$ consists of the classes of $Br(k)$ which have uniformly distributed invariants with values 0 or 1/2 such that the p-*local invariant is* 0 *whenever* p *does not split in* k.

Now we shall consider the Schur subgroup of $k = Q(\zeta_{q^n})$ $(\neq Q)$, where q is a prime number. Let p be an odd rational prime with $p \equiv 1 \pmod q$. Suppose $p = 1 + q^c d$ with $(q,d) = 1$. Let t be a primitive root modulo p. Let $\zeta = \zeta_{pq^n}$ be a primitive pq^n th root of unity. Then $\zeta^p = \zeta_{q^n}$ is a primitive q^n th root of unity. The Galois group $\mathfrak{G} = \mathfrak{G}(Q(\zeta)/k)$ is cyclic of order $p-1$: $\mathfrak{G} = \langle\sigma\rangle$, $\zeta^\sigma = \zeta^v$, where $v \equiv 1 \pmod{q^n}$ and $v \equiv t \pmod p$. Consider the following cyclotomic algebra A_p which is a cyclic algebra:

$$A_p = (\zeta_{q^n}, Q(\zeta)/k, \sigma) = \sum_{i=0}^{p-2} Q(\zeta)u^i \qquad \text{(direct sum)},$$

$$u^{p-1} = \zeta_{q^n}, \; u\zeta u^{-1} = \zeta^\sigma.$$

LEMMA 5.2. *If* y *is a rational prime other than* p, *the* y-*local index of* A_p *equals* 1. *If* $q \neq 2$ *or* $p \not\equiv -1 \pmod{q^n}$, *the* p-*local index of* A_p *equals* q^s, *where* $s = \mathrm{Min}\{n,c\}$. *If* $q = 2$ *and* $p \equiv -1 \pmod{2^n}$, *the* p-*local index of* A_p *equals* 1.

PROOF: If $\mathfrak{Q}$ is a finite prime of k which does not divide p, then $\mathfrak{Q}$ is non-ramified in K/k, and so $A_p \otimes_k k_{\mathfrak{Q}} \sim 1$. Every infinite prime $\mathfrak{P}_\infty$ of k is complex, and so $A_p \otimes_k k_{\mathfrak{P}_\infty} \sim 1$. Let $\mathfrak{P}$ be a prime of k dividing p. Denote by $W(K^{\mathfrak{P}})$ the finite subgroup of $(K^{\mathfrak{P}})^\times$ consisting of those roots of unity whose order are relatively prime to p. Let f be the smallest positive integer such that $p^f \equiv 1 \pmod{q^n}$. Denote by N the image of $W(K^{\mathfrak{P}})$ by the norm of $K^{\mathfrak{P}}$ over $k_{\mathfrak{P}}$. As $K^{\mathfrak{P}}/k_{\mathfrak{P}}$ is totally ramified, it follows that $W(K^{\mathfrak{P}}) = \langle\zeta_{p^f-1}\rangle$ and $N = \langle\zeta_{p^f-1}^{p-1}\rangle$. The $\mathfrak{P}$-index of A_p is the smallest positive integer m such that $\zeta_{q^n}^m \in N$. Recall that $p = 1 + q^c d$, $(q,d) = 1$. If $c \geq n$, then $s = n$ and $N = \langle\zeta_{p-1}^{p-1}\rangle = 1$. Consequently, $m = q^n = q^s$. Assume that $c < n$, and that $q \neq 2$ or $p \equiv -1 \pmod{q^n}$. Then we see easily that $f = q^{n-c}$, $p^f \not\equiv 1 \pmod{q^{n+1}}$ and $N = \langle \zeta_{\frac{p^f-1}{p-1}}^{\,p-1} \rangle = \langle \zeta_{q^{n-c}}^{\,z} \rangle$ where z is a certain integer relatively prime to q. Consequently, $m = q^c = q^s$. If $q = 2$ and $p \equiv -1 \pmod{2^n}$ then $f = 2$ and $N = \langle \zeta_{\frac{p^f-1}{p-1}}^{\,p-1} \rangle = \langle \zeta_{p+1} \rangle$. Since $p+1 \equiv 0 \pmod{2^n}$, it follows that $\zeta_{2^n} \in N$, and so $m = 1$.

THEOREM 5.3. (Benard-Schacher [4]). <u>Let</u> $k = \mathbb{Q}(\zeta_{q^n})$ <u>with</u> $q^n > 2$. <u>Then</u> $S(k) = S_{q^n}(k) \times S_2(k)$ <u>unless</u> $q = 2$, <u>in which case</u> $S(k) = S_{q^n}(k)$. $S_{q^n}(k)$ <u>is the Sylow</u> q-<u>subgroup of</u> $S(k)$ <u>and generated by the classes</u> $[A_p]$ <u>constructed above when</u> p <u>ranges over all primes</u> $p \equiv 1 \pmod{q}$. $S_2(k)$ <u>consists of those classes of</u> $Br(k)$ <u>which have uniformly distributed invariants with values</u> 0 <u>or</u> $1/2$ <u>such that the</u> p-<u>local invariants are</u> 0 <u>whenever</u> $p^a \not\equiv 1 \pmod{q^n}$ <u>for all odd integers</u> a.

PROOF: See [4].

ADDENDUM. After this paper was completed, the Schur subgroup $S(k)$ for an arbitrary subfield k of $\mathbb{Q}(\zeta_{q^c})$ has been characterized. In particular, every real subfield k of $\mathbb{Q}(\zeta_{q^c})$ has the property (M). The method of proof is entirely the same as in the case $k = \mathbb{Q}(\zeta_{q^c} + \zeta_{q^c}^{-1})$ or $k = \mathbb{Q}(\sqrt{q})$, $q \equiv 1 \pmod{4}$. If $[k:\mathbb{Q}]$ is odd, then by Corollary 4.3, k has the property (M). In particular we are done with the case $q \equiv 3 \pmod{4}$. Suppose that $q \equiv 1 \pmod{4}$ and $[k:\mathbb{Q}]$ is even. For a prime $p \neq 2$, set $K = \mathbb{Q}(\zeta_{q^c}, \zeta_p)$, $\mathcal{G}(K/\mathbb{Q}(\zeta_{q^c})) = \langle \theta \rangle$ and $\mathcal{G}(K/k(\zeta_p)) = \langle \varphi \rangle$. Define the cyclotomic algebra $B_p = (\varepsilon, K/k) = \sum_i \sum_j K u_\theta^i u_\varphi^j$ with relations $u_\theta u_\varphi = -u_\varphi u_\theta$, $u_\theta^{p-1} = u_\varphi^s = 1$, $s = [\mathbb{Q}(\zeta_{q^c}):k]$. For the prime 2, set $K = \mathbb{Q}(\zeta_{q^c}, \zeta_4)$, $\mathcal{G}(K/\mathbb{Q}(\zeta_{q^c})) = \langle \theta \rangle$ and $\mathcal{G}(K/k(\zeta_4)) = \langle \varphi \rangle$. Define the cyclotomic algebra $B_2^{\pm} = (\varepsilon, K/k) = \sum_i \sum_j K u_\theta^i u_\varphi^j$ with relations $u_\varphi u_\theta = \zeta_4 u_\theta u_\varphi$, $u_\varphi^s = \zeta_4^{s/2}$, $u_\theta^2 = \pm 1$. By virtue of these cyclotomic algebras, we conclude that k has the property (M). The real subfields of $\mathbb{Q}(\zeta_{2^c})$ have been already settled by Yamada [14]. If k is an imaginary subfield of $\mathbb{Q}(\zeta_{2^c})$ which does not contain roots of unity other than ± 1, then we see that $S(k) = S(\mathbb{Q}) \otimes_{\mathbb{Q}} k$. This is also true for the case $\mathbb{Q}(\zeta_{q^c})$, $q \neq 2$ by [4, Lemma 2].

REFERENCES

[1] A. A. Albert, Structure of algebras, Amer. Math. Soc., Providence, R.I., 1961.

[2] M. Benard, Quaternion constituents of group algebras, Proc. Amer. Math. Soc., 30 (1971), 217-219.

[3] M. Benard, The Schur subgroup, I, (to appear).

[4] M. Benard and M. M. Schacher, The Schur subgroup, II, (to appear).

[5] R. Brauer, On the representations of groups of finite order, Proc. Internat. Congress, Cambridge, 1950, Vol. 2, 33-36.

[6] R. Brauer, On the algebraic structure of group rings. J. Math. Soc. Japan, 3 (1951), 237-251.

[7] M. Deuring, Algebren, Springer, Berlin, 1935.

[8] K. L. Fields, On the Brauer-Speiser theorem, Bull. Amer. Math. Soc., 77 (1971), 223.

[9] K. L. Fields and I. N. Herstein, On the Schur subgroup of the Brauer group, J. Algebra, 20 (1972), 70-71.

[10] J.-M. Fontaine, Sur la décomposition des algebres de groupes, Ann. Sci. Ecole Norm. Sup., 4 (1971), 121-180.

[11] G. J. Janusz, The Schur index and roots of unity, (to appear).

[12] E. Witt, Die algebraische Struktur des Gruppenringes einer endlichen Gruppe über einem Zahlkörper, J. Reine Angew. Math., 190 (1952), 231-245.

[13] T. Yamada, Characterization of the simple components of the group algebras over the p-adic number field, J. Math. Soc. Japan, 23 (1971), 295-310.

[14] T. Yamada, Central simple algebras over totally real fields which appear in Q[G], Queen's University Preprint No. 1971-29.

[15] T. Yamada, The Schur subgroup, Queen's Papers, (to appear).

[16] H. J. Zassenhuas, The Theory of Groups, 2nd ed., Chelsea, New York, 1958.

ON THE EMBEDDING OF AN ORDER INTO A MAXIMAL ORDER

Hans Zassenhaus

Ohio State University

INTRODUCTION. At the Montreal Symposium on the application of number theory to numerical analysis, October, 1971, I gave a report "On the second round of the maximal order program" which describes the embedding of a $\mathbb{Z}$-order Λ into a maximal order. This report, the third round, develops a new method retaining just a few features of the previous round in order to produce <u>all</u> maximal orders over a given $\mathbb{Z}$-order Λ via appropriate algorithmic ways.

THE THIRD ROUND OF THE ORD MAX PROGRAM. The third round is guided by the idea that for large values of N (say $N > 10$) one must avoid working with sub-algebras with approximately N^2 basis elements as much as possible. Therefore one wants to split the problem into as many little pieces as possible. This is accomplished by a search for primitive idempotents.

Suppose R is a dedekind ring with quotient field F, $\mathfrak{p}$ a prime ideal of R, Λ an R-order (see [7]), E an element of Λ representing an idempotent residue class of Λ modulo $\mathfrak{p}\Lambda$. What advantage can be drawn from the knowledge of E which can be applied to the task of embedding Λ into a maximal order?

Informally, the development is as follows. Let S denote the half group $R\backslash\mathfrak{p}$ formed by the elements of R not in $\mathfrak{p}$ and let $S^{-1}R$ and $S^{-1}\Lambda$ denote the $\mathfrak{p}$-localizations of R and Λ respectively. The numerical advantage of localization consists in the existance of an $S^{-1}R$ basis for $S^{-1}\Lambda$ and a generator π for the maximal ideal $\mathfrak{p}S^{-1}\Lambda$ in $S^{-1}\Lambda$.

The construction is guided by the behavior of the $\mathfrak{p}$-adic completion $\Lambda_{\mathfrak{p}}$ of Λ whose elements w can be regarded as power series in π

$$w = \sum_{j=0}^{\infty} w_j \pi^j$$

where the coefficients w_j are taken from a fixed set of representatives of $\Lambda/p\Lambda$. In these terms, the existance of an idempotent residue class $E/p\Lambda$ in $\Lambda/p\Lambda$ implies the existance of an idempotent

$$E' = \sum_{j=0}^{\infty} x_j \pi^j$$

in the completion such that $E'/p\Lambda_p = E/p\Lambda_p$. E' gives rise to a suborder $E'\Lambda_p E'$ in Λ_p of smaller R_p-rank in which we work to search for larger orders. Specifically

LEMMA 0. <u>Let E_0 be an idempotent in an R_p-order Λ^* and let x be an R_p-order in $F_p(E_0\Lambda^*E_0)$. Then</u>

$$x + x\Lambda^*(1 - E_0) + (1 - E_0)\Lambda^*x + (1 - E_0)(\Lambda^* + \Lambda^* x \Lambda^*)(1 - E_0)$$

<u>is an R_p-order containing</u> Λ^*.

The proof of this lemma follows from the two-sided Pierce decomposition of Λ^*

$$\Lambda^* = E_0\Lambda^*E_0 + E_0\Lambda^*(1 - E_0) + (1 - E_0)\Lambda^*E_0 + (1 - E_0)\Lambda^*(1 - E_0)$$

and the fact that x contains $E_0\Lambda^*E_0$ and hence annihilates terms of decomposition involving $1 - E_0$. Of course the same result carries over to orders over any ring.

Computationally it is not possible to list all the coefficients of an idempotent E' in Λ_p (although one could develop a resursive relation). Instead we use a trancated version

$$E'' = \sum_{j=0}^{r} x_j \pi^j$$

for the construction of a p-extension of Λ.

These remarks should be kept in mind during the following formal exposition.

First, two definitions.

DEFINITION 1. The element E of Λ is said to be p-__primitive__ if the coset $E/p\Lambda$ is a primitive idempotent of $\Lambda/p\Lambda$.

DEFINITION 2. The element E of Λ is said to be p-__maximal__ if $0 \not\equiv E \equiv E^2$ (mod $p\Lambda$) and there is no R-order Λ^* containing Λ and an element E^* such that

$$0 \not\equiv E^* \equiv E^{*2} \pmod{p\Lambda^*},$$

$$(E^* \Lambda^* E^* + p\Lambda^*) \cap \Lambda = E\Lambda E + p\Lambda,$$

$$E\Lambda E + p\Lambda^* \subset E^* \Lambda^* E + p\Lambda^* .$$

LEMMA 3. (a) __If the element__ E __of__ Λ __is__ p-__maximal then in the__ p-__adic completion__ Λ_p __of__ Λ __there is an idempotent__ E_0^* __such that__

$$E_0^* \equiv E \pmod{p\Lambda_p},$$

$$E_0^* \Lambda_p E_0^*$$

__is a maximal__ R_p-__order in__ $E_0^*(F_p \Lambda_p)E_0^*$.

(b) __Conversely if__ Λ_p __contains an idempotent__ E_0 __such that__ $E_0 \Lambda_p E_0$ __is a maximal__ R_p-__order and__ E __is an element of__ Λ __such that__

$$0 \not\equiv E \equiv E^2 \pmod{p\Lambda},$$

$$E\Lambda_p E + p\Lambda_p = E_0 \Lambda_p E_0 + p\Lambda_p$$

<u>then</u> E <u>is</u> p-<u>maximal</u>.

PROOF OF (a): According to Zassenhaus 1954, there is an idempotent E_1 of Λ_p for which $E \equiv E_1 \pmod{p\Lambda_p}$. If $E_1\Lambda_pE_1$ were properly contained in an R_p-order Λ_0 of the same R_p-rank then we would have $\Lambda_0 = E_1\Lambda_0E_1$.

Furthermore, as pointed out in Lemma 0, Λ_p would be properly contained in the R_p-order

$$\Lambda_p^* = \Lambda_0 + \Lambda_0\Lambda_p(1 - E_1) + (1 - E_1)\Lambda_p\Lambda_0 + (1 - E_1)(\Lambda_p + \Lambda_p\Lambda_0\Lambda_p)(1 - E_1).$$

There would be a natural number ν for which

$$\Lambda_p^* \subseteq p^{-\nu}\Lambda_p ,$$

the intersection

$$\Lambda_p^* \cap p^{-\nu}\Lambda = \Lambda^*$$

would be an R-order properly containing Λ such that Λ_p^* would be the p-adic completion of Λ^*,

$$(E\Lambda^*E)_p = E_0\Lambda_p^*E_0 ,$$

$$(E\Lambda^*E + p\Lambda^*) \cap \Lambda = E\Lambda E + p\Lambda ,$$

$$E\Lambda E + p\Lambda^* \subset E\Lambda^*E + p\Lambda^* .$$

This contradicts the p-maximality of E.

PROOF OF (b): It follows from the assumption of (b) that both E and E_0 act as unit element of $E_0\Lambda_pE_0/p\Lambda_p$, hence

$$E \equiv E_0 \pmod{p\Lambda_p} .$$

There exist an idempotent E_1 of Λ_p such that

$$(E\Lambda E)_p = E_1 \Lambda_p E_1 ,$$

$$E_1 \equiv E_0 \pmod{p\Lambda_p} .$$

$E_1 \Lambda_p E_1$ is conjugate to $E_0 \Lambda_p E_0$ under the unit group of Λ_p according to Zassenhaus 1954, thus $E_1 \Lambda_p E_1$ is a maximal R_p - order.

THEOREM 4. <u>The element</u> E <u>of</u> Λ <u>is</u> p - <u>maximal and</u> p - <u>primitive if and only if</u>

(i) $0 \not\equiv E \equiv E^2 \pmod{p\Lambda}$

(ii) <u>the factor ring of</u> $(E\Lambda E + p\Lambda)/p\Lambda$ <u>over the radical ideal</u> $J_p(E\Lambda E + p\Lambda)/p\Lambda$ <u>is a division ring</u>

(iii) <u>there is a natural number</u> e <u>such that</u>

$$(J_p(E\Lambda E + p\Lambda))^{e-1} \not\subseteq p\Lambda ,$$

$$(J_p(E\Lambda E + p\Lambda))^{e} + p^2\Lambda + (1 - E)p\Lambda + (p\Lambda)(1 - E) = p\Lambda .$$

PROOF: Firstly let E a p - maximal and p - primitive element of Λ. It follows from the definition of $J_p(E\Lambda E + p\Lambda)$ as the largest ideal of $E\Lambda E + p\Lambda$ with nilpotent factor ring over $p\Lambda$, that the factor ring of $E\Lambda E + p\Lambda$ over $J_p(E\Lambda E + p\Lambda)$ is semisimple. It follows from Zassenhaus 1954, that $E/J_p(E\Lambda E + p\Lambda)$ is a primitive idempotent. Since it is also the identity element of $(E\Lambda E + p\Lambda)/J_p(E\Lambda E + p\Lambda)$ it follows from the McLagan-Wedderburn structure theorems that the factor ring of $E\Lambda E + p\Lambda$ over $J_p(E\Lambda E + p\Lambda)$ is a division ring.

According to Zassenhaus 1954, there is an idempotent E^* of the p - adic completion Λ_p of Λ such that $E^* \equiv E \pmod{p\Lambda_p}$. Since E is p - maximal it follows from Lemma 3 that $E^* \Lambda_p E^*$ is a maximal R_p - order. Hence it follows the Benz-Zassenhaus Theorem (see [1]) that 4.1 holds for some natural number e.

Conversely, let E be an element of Λ for which $0 \not\equiv E \equiv E^2 \pmod{p\Lambda}$, let $(E\Lambda E + p\Lambda)/J_p(E\Lambda E + p\Lambda)$ be a division ring and let 4.1 be satisfied for some natural number e.

According to Zassenhaus 1954, there is an idempotent E^* of Λ_p such that $E^* \equiv E \pmod{p\Lambda^*}$. Hence the factor ring of the R_p-order $E^*\Lambda_p E^*$ over its Jacobson radical $J(E^*\Lambda_p E^*)$ is isomorphic to $(E\Lambda E + p\Lambda)/J_p(E\Lambda E + p\Lambda)$ so that is is a division algebra over R_p/pR_p. Because of 4.1 we have $J(E^*\Lambda_p E^*)^e = pE^*\Lambda_p E^*$ so that according to Benz-Zassenhaus the order $E^*\Lambda_p E^*$ is hereditary. It is maximal because the factor ring of $E^*\Lambda_p E^*$ over its radical is a division ring. Hence E is p-maximal and p-primitive.

COROLLARY 5. _If_ E _is_ p-_primitive, but not_ p-_maximal in_ Λ, _then there is a natural number_ $e > 1$ _for which_

$$(J_p(E\Lambda E + p\Lambda))^{e-1} \not\subset p\Lambda,$$

$$(J_p(E\Lambda E + p\Lambda))^{e} + p^2\Lambda + (1 - E)p\Lambda + p\Lambda(1 - E) \subset p\Lambda$$

and consequently Λ _is properly contained in the_ R-_order_

$$\Lambda^* = \Lambda + X + X\Lambda(1 - E) + (1 - E)\Lambda X + (1 - E)\Lambda X \Lambda(1 - E)$$

where

$$X = p^{-1}(J_p(E\Lambda E + p\Lambda))^{e-1}$$

PROOF: This is modelled along the lines of the computations needed in the proof of Proposition 0.

It follows from the assumptions that

$$p\Lambda \subseteq XJ_p(E\Lambda E + p\Lambda) + p\Lambda \subset E\Lambda E + p\Lambda,$$

hence

$$XJ_p(E\Lambda E + p\Lambda) \subseteq J_p(E\Lambda E + p\Lambda)$$

because of the p-primitivity of

$$XX = p^{-1}XJ_p(E\Lambda E + p\Lambda)^{e-1}$$
$$\subseteq p^{-1}J_p(E\Lambda E + p\Lambda)^{e-1} = X .$$

Furthermore

$$X(1 - E) = p^{-1}(J_p(E\Lambda E + p\Lambda)^{e-1}(1 - E)$$
$$\subseteq p^{-1}(E\Lambda E(1 - E)) + \Lambda \subseteq \Lambda,$$

similarly

$$(1 - E)X \subseteq \Lambda .$$

Finally

$$J_p(E\Lambda E + p\Lambda) = ((J_p(E\Lambda E + p\Lambda)) \cap E\Lambda E) + p\Lambda$$

$$EX = Ep^{-1}(J_p(E\Lambda E + p\Lambda))^{e-1}$$
$$\subseteq \Lambda + p^{-1}J_p(E\Lambda E + p\Lambda)^{e-1}$$
$$= \Lambda + X$$

and similarly

$$XE \subseteq \Lambda + X .$$

Using these set theoretical relations we prove similarly as above that $\Lambda^*\Lambda^* \subseteq \Lambda^*$.

These results provide an opportunity of carrying out the major part of the subsequent algorithm inside the "little pieces" $E_i p^{-\rho}\Lambda E_i$, $(1 \leq i \leq r)$: Only if all of the E_i's are p-maximal and p-primitive it would be necessary to determine the interrelation of the p-maximal extension Λ^{**} of Λ with the connecting modules $E_i p^{-\rho}\Lambda E_k$. At that stage one can compute the discriminant of Λ^{**} without computing the $E_i \Lambda^{**} E_k$.

We do not give the algorithm in this streamlined form, but expound it in a way appropriate for smaller dimensions.

COROLLARY 6 OF THEOREM 4. *Let* E *be* $\mathfrak{p}$-*primitive and* $\mathfrak{p}$-*maximal in* Λ, *let* E' *be some element of* Λ *for which*

$$E'^2 \equiv E' \pmod{\mathfrak{p}\Lambda},$$

let ρ *be a natural number for which*

$$E^2 \equiv E \pmod{\mathfrak{p}^\rho\Lambda}$$
$$E'^2 \equiv E' \pmod{\mathfrak{p}^\rho\Lambda}$$
$$E\Lambda E'\Lambda E \not\subseteq \mathfrak{p}^\rho\Lambda,$$

let N *be the preimage in* Λ *of the radical ideal of* $E\Lambda E + \mathfrak{p}^\rho\Lambda$, *let* e *be the natural number for which* $N^e \subseteq \mathfrak{p}\Lambda$, $N^{e-1} \not\subseteq \mathfrak{p}\Lambda$, *then we have*

$$E\Lambda E'\Lambda E = N^\sigma + \mathfrak{p}^\rho\Lambda$$

for some non-negative integer σ, *and there is the* R-*order*

$$\Lambda' = \Lambda + E'\Lambda(\mathfrak{p}^{-1}N^{e-1})^\sigma + (\mathfrak{p}^{-1}N^{e-1})^\sigma\Lambda E' + E'\Lambda(\mathfrak{p}^{-1}N^{e-1})^\sigma\Lambda E'$$

containing Λ *such that* $\mathfrak{p}^\rho\Lambda' \subseteq \Lambda$, *the composition length of the* R-*module* Λ'/Λ *is at least* σ *and that*

$$E'\Lambda(\mathfrak{p}^{-1}N^{e-1})^\sigma\Lambda E'$$

contains an element E'' *such that*

$$0 \not\equiv E''^2 \equiv E'E' \equiv E''E' \equiv E'' \pmod{\mathfrak{p}^\rho\Lambda'}.$$

PROOF: This follows upon $\mathfrak{p}$-adic analysis of the situation.

Before we enter the description of the embedding algorithm let us define two elements E, E' of an R-order Λ^* of $F\Lambda$ to be $\mathfrak{p}$-*equivalent* if

$$0 \not\equiv E \equiv E^2 \pmod{\mathfrak{p}\Lambda^*}, \quad 0 \not\equiv E' \equiv (E')^2 \pmod{\mathfrak{p}\Lambda^*}$$

and for some R-order Λ^{**} of $F\Lambda$ containing Λ^* one can find an element X of Λ^{**} for which $X/\mathfrak{p}\Lambda^{**}$ is contained in the unit group of $\Lambda^{**}/\mathfrak{p}\Lambda^{**}$ such that $XE \equiv E'X \pmod{\mathfrak{p}\Lambda^{**}}$. Again upon $\mathfrak{p}$-adic analysis of the situation one sees that $\mathfrak{p}$-equivalence is reflexive, symmetric and transitive.

At this point, let us describe the ensuing embedding algorithm in a more general setting.

Given a dedekind ring R with quotient field F and with a prime ideal $\mathfrak{p}$ with finite residue class field $k = R/\mathfrak{p}$. Also given an R-order with contribution $\mathfrak{p}^{\nu}$ to the discriminant ideal of Λ over R. We want to embed Λ into a $\mathfrak{p}$-maximal order of $F\Lambda$.

In case $\nu = 0$ or 1, Λ is already $\mathfrak{p}$-maximal. Now let $\nu \geq 2$.

Let p the characteristic of k, let $q = p^{N_0}$ the number of elements of k.

We consider an order Λ^* of $F\Lambda$ containing Λ such that the index of Λ^* over Λ is equal to q^{ν_1}. The contribution of $\mathfrak{p}$ to the discriminant ideal of Λ^* is equal to $\mathfrak{p}^{\nu-2\nu_1}$.

There are several ways to present Λ^* in computational terms. For example, one may give an exponent $\nu_2 \geq 0$ such that $\Lambda^* \subseteq \mathfrak{p}^{-\nu_2}\Lambda$, $\Lambda^* \not\subseteq \mathfrak{p}^{-\nu_2+1}\Lambda$, and one may describe an element x of Λ^* in terms of a given R-basis B of Λ in the form

$$x = \pi^{-\nu_2} \sum_{b \in B} \lambda_{xb}\, b$$

with λ_{xb} in R. If R is not principal, one may need to go to the $\mathfrak{p}$-localization for this.

Furthermore we assume that a natural number $\rho > \nu/2$ and certain elements $E_1, \dots, E_r$ of Λ^* are known such that for $1 \leq i, k \leq r$,

$$E_iE_k \equiv \begin{cases} E_i \not\equiv 0 & \text{if } i = k \\ 0 & \text{if } i \neq k \end{cases} \pmod{p^\rho \Lambda^*}$$

and $E_1 + E_2 + \dots + E_r = 1$, that there are non negative integers $r_0, s_0, r_{01}, \dots r_{0s_0}$ for which

$$r \geq r_0 = r_{01} + \dots + r_{0s_0}$$

$$r_{0j} > 0 \quad \text{if } 1 \leq j \leq s_0 ,$$

the E_i's are p-maximal and p-primitive if $1 \leq i \leq r_0$, also if $0 < i' < k' < s_0$,

$$\sum_{h<i'} r_{0h} < i \leq \sum_{h\leq i'} r_{0h} \leq \sum_{h<k'} r_{0h} < k \leq \sum_{h\leq k'} r_{0h}$$

then E_i and E_k are not p-equivalent but on the other hand, in case $0 < i' = k' \leq s_0$, and $\sum_{h<i'} r_{0h} < i$ and $k \leq \sum_{h<i'} r_{0h}$, E_i and E_k are p-equivalent so that $E_i \Lambda^* E_k \not\subseteq p^\rho \Lambda^*$. We assume that there are natural numbers e_i, $1 \leq i \leq r_0$ for which $N_i^{e_i - 1} \neq 0$ and $e_i = e_k$ in case E_i and E_k are p-equivalent, $1 \leq i < k \leq r_0$, that there are elements ξ_i of Λ^* known such that $E_i \xi_i \equiv \xi_i E_i \equiv \xi_i \pmod{p^\rho \Lambda}$ and $k_i = E_i R(\xi_i)/p\Lambda^*$, is an extension of degree μ_i of $E_i R/p\Lambda^*$, that there are elements $n_{i1}, \dots, n_{id_i}$, of $E_i \Lambda^* E_i$ known such that the cosets modulo $p\Lambda^*$ of these elements are linearly independent over k and the module $N_i = \sum_{j=1}^{d_i} n_{ij} R/p\Lambda^*$ is a nilpotnet ideal of $\Lambda_i^* = E_i \Lambda^* E_i / p\Lambda^*$, that $\Lambda_i^* = k_i + N_i$ if $1 \leq i \leq r_0$, that there is a subset B_i of $E_i \Lambda^* E_i$ such that $B_i/(p\Lambda^* + \sum_{j=1}^{d_i} n_{ij} R)$ is a k_i-left basis of Λ_i^*

modulo N_i for which

$$E_i \in B_i ,$$

$$b\,b' \equiv \sum_{b'' \in B_i} \lambda_{b''}^{(b,b')}\, b'' \pmod{N_i}$$

$$b\,\xi_i \equiv \sum_{b'' \in B_i} \lambda_{b''}^{(i)}\, b'' \pmod{N_i}$$

$$(\lambda_{b''}^{(b,b')},\ \lambda_{b''}^{(i)} \in k_i ,\ b,\ b',\ b'' \in B_i) ,$$

that there is some natural number r_1, such that $r_0 < r_1 \leq r + 1$ and for all indices satisfying $r_1 \leq i \leq r$ there is a subset B_i' of B_i whose cosets modulo $\mathfrak{p}\Lambda^*$ form a k_i-basis of the centralizer of k_i in the factor ring Λ_i^*/N_i, and that there are natural numbers r_2 such that $r_0 < r_1 \leq r + 1$ and for all indices satisfying $r_2 \leq i$, $r_1 \leq i \leq r$, it is known that E_i is $\mathfrak{p}$-primitive so that $B_i = B_i'$ consists only of E_i.

To start with set

$$\Lambda = \Lambda^*$$

$$E_1 = 1 ,$$

$$r = 1 ,\ r_0 = 0 ,\ r_1 = 1 ,\ r_2 = 2 ,\ d_1 = 0 ,\ \mu_1 = \rho_1 = 1 ,$$

$$k_1 = k ,\ N_1 = 0 ,\ B_1 = B_1' = B ,$$

$$b\,b' = \sum \varkappa_{b''}^{(b,b')}\, b'' ,$$

$$\varkappa_{b''}^{(b,b')} \in R ,$$

$$\lambda_{b''}^{(b,b')} = \varkappa_{b''}^{(b,b')}/\mathfrak{p}$$

$$(b,b',b'' \in B) .$$

At each stage of the embedding algorithm we proceed as follows,

I. If $\nu - 2\nu_1 \leq 1$ then Λ^* is p-maximal and we are done.

Let $\nu - 2\nu_1 > 1$. It follows from a theorem of Dickson that there is a decomposition of $\Lambda_i^* = E_i\Lambda^*E_i/p\Lambda^*$ of the form

$$\Lambda_i^* = J_i^* + T_i$$

when J_i^* is the radical of Λ_i^* and therefore J_i^* contains N_i and where T_i is a semi-simple subring of Λ_i^* containing R_i.

II. $r_0 + 1 < r_1$. In this case find a solution basis B_{r_1-1} modulo N_{r_1-1} of the system of linear equations

$$\sum_{b\in B_{r_1-1}} \xi_{r_1-1}\,\eta(b)b \equiv \sum_{b\in B_{r_1-1}} \eta(b)b\,\xi_{r_1-1} \pmod{N_{r_1-1}}$$

and exchange a suitable subset of B_{r_1-1} for B'_{r_1-1} such that a new basis $\hat{B}_{r_1}/N_{r_1-1}$ of $\Lambda^*_{r_1-1}/N_{r_1-1}$ will be obtained. Now replace $\hat{B}_{r_1}$ by B_{r_1}, r_1 by r_1-1.

If $r_0 + 1 = r_1 < r_2$, $r_1 \leq r$, $r_1 \leq i \leq r$, $i < r_2$, $b \in B_i'$ then the first power of b, say $b^{\mu(b)}$, that modulo N_i is linearly dependent over k_i on the previous powers E_i, b, ..., $b^{\mu(b)-1}$ determines uniquely the monic polynomial $f_b(t)$ of $k_i[t]$ such that

$$f_b(b/N_i) = 0 .$$

III. $r_0 + 1 = r_1 < r_2$, $r_1 \leq r$, and for some index i satisfying $r_1 \leq i \leq r$, $i < r_2$ we find some element b of B_i for which $f_b(t)$ is divisible by at least two distinct monic irreducible polynomials other than t, say

$$f_b(t) = t^{\varkappa_0} \prod_{j=1}^{\lambda} f_j(t)^{\varkappa_j}, \quad \lambda > 1, \quad \varkappa_0 \geq 0,$$

$\varkappa_j > 0$, $f_j(t)$ monic and irreducible of $k_i[t]$, $(1 \leq j \leq \lambda)$.

Hence we can determine polynomials $A_j(t)$ $(0 \leq j \leq \lambda)$ of $k_i[t]$

$$1 = A_0(t) \prod_{j=1}^{\lambda} f_j(t)^{\varkappa_j} + \sum_{j=1}^{\lambda} A_j(t) t^{\varkappa_0} \prod_{\substack{\tau=1 \\ \tau \neq j}} f_\tau(t)^{\varkappa_\tau}.$$

Then

$$E_i = \sum_{j=0}^{\lambda} E_{ij},$$

where

$$0 \neq E_{ij} = A_j(b)\, b^{\varkappa_0} \prod_{\substack{i=1 \\ i \neq j}}^{\lambda} f_i(b)^{\varkappa_i} \quad \text{for } 0 < j \leq \lambda$$

and

$$E_{i0} = A_0(b) \prod_{j=1}^{\lambda} f_i(b)^{\varkappa_i}$$

$$E_{ij} E_{ij'} \equiv \begin{cases} E_{ij} & \text{if } j = j' \\ 0 & \text{otherwise} \end{cases} \pmod{N_i}.$$

Solve the conditions

$$E_{ij} \in E_i \Lambda^* E_i,$$

$$E'_{ij} \equiv E_{ij} \pmod{N_i},$$

$$E'_{ij} E'_{ij'} \equiv \begin{cases} E'_{ij} & \text{if } j = j' \\ 0 & \text{otherwise} \end{cases} \pmod{N_i}$$

$$E_i = \sum_{j=0}^{\lambda} E'_{ij} .$$

Replace the E_{ij} by the E'_{ij}, now form the

$$n_{ijk} = E_{ij} n_{ij} E_{ij} .$$

Drop those n_{ijk}'s whose cosets modulo $p\Lambda^*$ are linearly dependent over k on the preceding ones. Renumber those remaining from 1 to $\bar{d}_{ij}$. Form the finite field $k_{ij} = E_{ij} k_i E_{ij} / p\Lambda^*$; form $B_{ij} = E_{ij} B_{ij} E_{ij}$, $B'_{ij} = E_{ij} B'_{ij} E_{ij}$. Order B_{ij} in such a way that the elements of B'_{ij} come first and $E_{ij} E_i E_{ij}$ is the first element. Drop those elements of B_{ij} whose cosets modulo $p\Lambda^*$ are linearly dependent over k_{ij} $(j = 1, 2, \ldots, \lambda$; in case $E_{i0} \not\equiv 0 \pmod{N_i}$ also $j = 0)$. Now renumber r, etc. so that the E_{ij}, n_{ijk}, B_{ij}, B'_{ij} together take the place of E_i, n_{ik}, B_i, B'_i.

IV. $r_0 + 1 = r_1 < r_2$, $r_1 \leq r$ and for some index i satssfying $r_1 \leq i \leq r$, $i < r_2$ we find some element b of B'_i for which $f_b(t)$ is divisible by some non-linear monic irreducible polynomial $g(t)$ of $k_i[t]$.

In this case, form

$$b' \equiv f/g\ (b) \pmod{p\Lambda^*}, \quad k'_i = k_i(b'/p\Lambda^*), \quad k'_i = k(\xi'_i) .$$

Replace k_i by k'_i, ξ_i by ξ'_i. Remove from the ordered set B_i (with E_i as first element) those elements that modulo $p\Lambda^*$ are linearly dependent over k'_i; interchange i and $r_1 - 1$; replace r_2 by $r_1 - 1$; continue with II, III.

V. $r_0 + 1 = r_1 < r_2$, $r_1 \leq r$ and for some index i satisfying $r_1 \leq i \leq r$, $i < r_2$ neither III nor IV can be applied. In that case k_i is a maximal abelian subring of T_i. Since k_i is a field it follows that T_i is simple. If $\mu_i > 1$, then

$$\mu_i = \prod_{j=1}^{\tau} p_j^{\sigma_j}$$

where $p_1, \ldots, p_\tau$ distinct prime numbers and $\sigma_j > 0$ $(1 \leq j \leq \tau)$. If one of the linear congruences

$$x \xi_i^{q^{\mu_i/p_j}} \equiv \xi_i x \pmod{N_i}$$

is satisfied by some element x of $E_i \Lambda^* E_i$ then from x and ξ_i, k_i one can construct explicitly a decomposition of $E_i / p\Lambda^*$ into the sum of several orthogonal idempotents modulo N_i (see Zassenhaus 1954, p. 55) say $E_i \equiv \sum E_{ij}$ $(\text{mod } N_i)$. Now proceed as in III.

If none of the linear congruences stated above has a solution in $E_i \Lambda^* E_i$, then $k_i = T_i$. This also happens if $\mu = 1$. In this case, for every element b of B_i there is some element $\alpha(b)$ of k_i such that $b/N_i - \alpha(b)$ is nilpotent and $\alpha(b)$ is a root of the minimal equation $m_b(b/N_i)$ of b modulo N_i over k.

There are only finitely many solutions of the equation $m_b(\xi) = 0$ in k_i. Precisely one of them has the property that a power of $b - \xi$ is in N_i and that ξ is equal to $\alpha(b)$. Now replace b by $b - \alpha(b)$ in case $b \neq E_i$. Join the new elements $\xi_i^{*} b$ $(0 \leq \varkappa < \mu_i,\ E_i \neq b \in B_i)$ with the set of the n_{ik}, replace B_i by $\{E_i\}$, interchange i and $r_2 - 1$, replace r_2 by $r_2 - 1$.

VI. $r_0 + 1 = r_1 < r_2 \leq r$. Determine the natural number e for which $N_r^e = 0$, $N_r^{e-1} \neq 0$. If

$$(pE_r \Lambda^* E_r + \sum_{j=1}^{d_r} n_{ij} R)^{-e} = p^2 \Lambda^* + pE_r \Lambda^* E_r$$

then E_r is p-maximal. Otherwise replace Λ^* by the order Λ^{**} generated by Λ^* and $p^{-1}E(pE_r\Lambda^*E_r + \sum_{j=1}^{d_r} n_{ij}R)^e$ according to Corollary 5. Because of the assumed p-maximality of the E_i' with $i \leq r_0$ the corresponding factor rings Λ_i^* will not be afftected. All other factor rings $\Lambda_j^*\,(j > r_0)$ may be affected by the transition from Λ^* to Λ^{**} with concomitant revisions of the n_{jk}, B_j, B_j' as explained earlier. Of course ν_1 will also be increased.

In case E_r is p-maximal, one invokes Corollary 6 for those i's for which $r_0 < i < r$, and $E_r\Lambda^*E_i\Lambda^*E_r \not\subseteq p^\rho\Lambda^{**}$. It follows that $E_r\Lambda^*E_i\Lambda^*E_r = N^{\sigma_i} + p^\rho\Lambda$ for some non negative integer σ_i with

$$N = J_p(E_r\Lambda^*E_r + p^\rho\Lambda^*) = \sum_{j=1}^{d_r} n_{rj}R + pE_r\Lambda^*E_r + p^\rho\Lambda^* .$$

We form the R-order Λ^{**} generated by Λ^* and the modules

$$E_i\Lambda(p^{-1}N^{e-1})^{\sigma_i},\ (p^{-1}N^{e-1})^{\sigma_i}\Lambda E_i,\ E_i\Lambda(p^{-1}N^{e-1})^{\sigma_i}\Lambda E_i .$$

Again, this construction will not affect the factor rings Λ_h^* if either $h \leq r_0$ or $h = r$. All other factor rings Λ_j^* may be affected by the transition from Λ^* to Λ^{**} with concomitant revisions of the n_{jk}, B_j, B_j' as had been explained earlier. Of course, ν_1 will be increased.

If at some stage of the algorithm none of the factor rings Λ_j^* $(r_0 < j < r)$ is affected then we gather the index r and the set of indices i satisfying $r_0 < i < r$, $E_r\Lambda^*E_i\Lambda^*E_r \subseteq p^\rho\Lambda^*$ into a finite set of r_{0,s_0+1} indices which will be placed in the position

$$r_0 + 1,\ \ldots,\ r_0 + r_{0,s_0+1}$$

shifting the other indices beyond $r_0 + r_{0,s_0+1}$. Let $e_i = e$. If

$r_0 < i \leq r_0 + r_{0,s_0+1}$, change s_0 to s_0+1.

VII. $r_0 = r$. Gather all pairs of indices i, k for which

$$0 < i' \leq s_0, \quad 1 + \sum_{h<i'} r_{0h} = i < k \leq \sum_{h\leq i'} r_{0h}, \quad E_i \Lambda^* E_k \Lambda^* E_i \subseteq p\Lambda^* .$$

Since ρ was chosen sufficiently large it follows that

$$E_i \Lambda^* E_k \Lambda^* E_i \not\subseteq p^\rho \Lambda^* ,$$
$$E_i \Lambda^* E_k \Lambda^* E_i = p^\rho \Lambda^* + J_p(E_i \Lambda^* E_i + p^\rho \Lambda^*)^{\sigma_{ik}}$$

with positive exponent σ_{ik}. We form the order generated by Λ^* and by all

$$E_k \Lambda (p^{-1} J_\Lambda (E_i \Lambda^* E_i + p^\rho \Lambda^*)^{e-1})^{\sigma_{ik}} \Lambda E_k .$$

It is p-maximal and contains Λ.

If one wants to compute <u>all</u> p-maximal orders containing Λ then according to the analysis given in VI, it suffices to do this in the case that 1 is p-primitive. The techniques of the Benz-Zassenhaus Theorem can then be applied to carry out the construction.

REFERENCES

[1] H. Benz and H. Zassenhaus, Untersuchungen zur Arithmetik in lokalen einfachen Algebren, 1969-72 (unpublished).

[2] E. R. Berlekamp, On the factorization of polynomials over finite fields, Bell Telephone Laboratories, Inc., Murrary Hill, N. J. (1967).

[3] Horst Kempfert, On the factorization of polynomials, Journal of Number Theory 1 (1969), 116-120.

[4] Hans Zassenhaus, Über eine Verallgemeinerung des Henselschen Lemmas, Arch. Math. 5 (1954), 317-325.

[5] ____________, Über die Fundamental konstruktionen der endlichen Körpertheorie, Jahresbericht der DMV 70 (1968), 177-181.

[6] ____________, On Hensel factorization I, Journal of Number Theory 1 (1969), 291-311.

[7] ____________, On the Second Round of the Maximal Order Program: Applications of Number Theory to Numerical Analysis, Academic Press, New York and London 1972, 389-432.

INDEX